T0142069

Studies in Systems, Decision and Control

Volume 14

Series editor

Janusz Kacprzyk, Polish Academy of Sciences, Warsaw, Poland
e-mail: kacprzyk@ibspan.waw.pl

About this Series

The series "Studies in Systems, Decision and Control" (SSDC) covers both new developments and advances, as well as the state of the art, in the various areas of broadly perceived systems, decision making and control- quickly, up to date and with a high quality. The intent is to cover the theory, applications, and perspectives on the state of the art and future developments relevant to systems, decision making, control, complex processes and related areas, as embedded in the fields of engineering, computer science, physics, economics, social and life sciences, as well as the paradigms and methodologies behind them. The series contains monographs, textbooks, lecture notes and edited volumes in systems, decision making and control spanning the areas of Cyber-Physical Systems, Autonomous Systems, Sensor Networks, Control Systems, Energy Systems, Automotive Systems, Biological Systems, Vehicular Networking and Connected Vehicles, Aerospace Systems, Automation, Manufacturing, Smart Grids, Nonlinear Systems, Power Systems, Robotics, Social Systems, Economic Systems and other. Of particular value to both the contributors and the readership are the short publication timeframe and the world-wide distribution and exposure which enable both a wide and rapid dissemination of research output.

More information about this series at http://www.springer.com/series/13304

Jaime Nava · Vladik Kreinovich

Algorithmic Aspects of Analysis, Prediction, and Control in Science and Engineering

An Approach Based on Symmetry and Similarity

 Springer

Jaime Nava
Department of Computer Science
University of Texas at El Paso
El Paso, Texas
USA

Vladik Kreinovich
Department of Computer Science
University of Texas at El Paso
El Paso, Texas
USA

ISSN 2198-4182
ISBN 978-3-662-51159-6
DOI 10.1007/978-3-662-44955-4

ISSN 2198-4190   (electronic)
ISBN 978-3-662-44955-4   (eBook)

Springer Heidelberg New York Dordrecht London

Printed on acid-free paper

Springer is part of Springer Science+Business Media (www.springer.com)

Preface

Algorithms are extremely important in science and engineering. One of the main objectives of science is to *predict* future events; this usually requires sophisticated algorithms. Once we are able to predict future events, a natural next step is to *influence* these events, i.e., to *control* the corresponding systems; control also usually requires complex algorithms. To be able to predict and control a system, we need to have a good *description* of this system, so that we can use this description to *analyze* the system's behavior and extract the desired prediction and control algorithms from this analysis.

A typical prediction is based on the fact that we observed similar situations in the past; we know the outcomes of these past situations, and we expect that the future outcome of the current situation will be similar to these past observed outcomes. In mathematical terms, similarity corresponds to *symmetry*, and similarity of outcomes – to *invariance*.

Because symmetries are ubiquitous and useful, we will show how symmetries can be used in all classes of algorithmic problems of sciences and engineering: from analysis to prediction to control. Specifically, we show how the approach based on symmetry and similarity can be used in the analysis of real-life systems, in the algorithmics of prediction, and in the algorithmics of control.

Our thanks to Dr. Benoit Bagot, Dr. Art Duval, Dr. Larry Ellzey, Dr. Yuri Gurevich, Dr. Luc Longpré, Dr. James Salvador, and Dr. Patricia Nava for their help, valuable suggestions, and inspiring ideas.

Thanks as well to all the participants of the 65th Southwest Regional Meeting of the American Chemical Society (El Paso, Texas, November 4–7, 2009), the 6th Joint UTEP/NMSU Workshop on Mathematics, Computer Science, and Computational Sciences (El Paso, Texas, November 7, 2009), the 14th GAMM-IMACS International Symposium on Scientific Computing, Computer Arithmetic and Validated Numerics SCAN'2010 (Lyon, France, September 27–30, 2010), and the Annual Conference of the North American Fuzzy Information Processing Society NAFIPS'2011 (El Paso, Texas, March 18-20, 2011) for valuable discussions.

This work was supported in part by the National Science Foundation grants HRD-0734825 and DUE-0926721, by Grant 1 T36 GM078000-01 from the National Institutes of Health, by Grant MSM 6198898701 from MŠMT of Czech Republic, and by Grant 5015 "Application of fuzzy logic with operators in the knowledge based systems" from the Science and Technology Centre in Ukraine (STCU), funded by European Union.

El Paso, Texas,   *Jaime Nava*
August 2014   *Vladik Kreinovich*

Contents

Chapter 1
Introduction: Symmetries and Similarities as a Methodology for Algorithmics of Analysis, Prediction, and Control in Science and Engineering

Need for Analysis, Prediction, and Control in Science and Engineering. One of the main objectives of science is to *predict* future events. Once we are able to predict future events, a natural next step is to *influence* these events, i.e., to *control* the corresponding systems. In this step, we should select a control that leads to the best possible result.

To be able to predict and control a system, we need to have a good *description* of this system, so that we can use this description to *analyze* the system's behavior and extract the desired prediction and control algorithms from this analysis.

Symmetry and Similarity: A Fundamental Property of the Physical World. As we have just mentioned, one of the main objectives of science is prediction. What is the usual basis for prediction?

A typical prediction is based on the fact that we observed similar situations in the past; we know the outcomes of these past situations, and we expect that the future outcome of the current situation will be similar to these past observed outcomes.

In mathematical terms, similarity corresponds to *symmetry*, and similarity of outcomes – to *invariance*.

Example: Geometric Symmetries and Related Similarities. For example, we dropped the ball, it fell down. We conclude that if we drop it at a different location and/or at a different orientation, it will also fall down. Why – because we believe that the corresponding process is invariant with respect to shifts, rotations, etc.

In this example, we used *geometric* symmetries, i.e., symmetries like shift, rotation, etc., that have a direct geometric meaning.

Example: Discrete Geometric Symmetries and Related Similarities. In the above example, the corresponding symmetries form a *continuous* family. In some other situations, we only have a *discrete* set of geometric symmetries.

In many practical situations, molecules can be obtained from a symmetric "template" molecule like benzene C_6H_6 by replacing some of its hydrogen atoms with *ligands* (other atoms or atom groups). For example, a sphere is invariant with respect to arbitrary rotations, but molecules such as benzene or cubane C_8H_8 are invariant with respect to certain rotations. For benzene, rotation by $60°$ transforms the first

© Springer-Verlag Berlin Heidelberg 2015
J. Nava and V. Kreinovich, *Algorithmic Aspects of Analysis, Prediction, and Control in Science and Engineering*, Studies in Systems, Decision and Control 14, DOI: 10.1007/978-3-662-44955-4_1

atom into the second one, the second into the third one, etc. In general, for every two atoms, we can find a rotation that moves the first atom into the position of the second one while keeping the molecule configuration intact.

Such a rotation does not change the chemical properties of a molecule – and hence, does not change the values of any numerical property of the substance. This symmetry helps us to predict, e.g., properties of *monosubstituted* molecules, i.e., molecules in which a ligand is placed at one of the locations. All the monosubstituted molecules can be obtained from each other by rotation; we can therefore conclude that all these molecules have the same values of all the numerical quantities.

More General Symmetries and Related Similarities. Symmetries can go beyond simple geometric transformations. For example, the current simplified model of an atom, in which electrons rotate around a nucleus, was originally motivated by an analogy with a Solar system, in which planets rotate around the Sun. The operation that transforms the Solar system into an atom has a geometric aspect: it simply scales down all the distances. However, this transformation goes beyond a simple geometric transformation, because in addition to changing distances, we also change masses, velocities, replace masses with electric charges, etc.

Basic Symmetries: Scaling and Shift. Let us start with the *basic* symmetries, i.e., symmetries directly related to the fact that in order to understand a real-life phenomenon, we must perform appropriate measurements.

As a result of each measurement, we get a numerical value of a physical quantity. Numerical values depend on the *measuring unit*. If we use a new unit which is λ times smaller, numerical values are multiplied by λ: $x \to \lambda \cdot x$. For example, x meters $= 100 \cdot x$ cm. The transformation $x \to \lambda \cdot x$ is usually called *scaling*.

Another possibility is to change the starting point. For example, instead of measuring time from year 0, we can start measuring it from some more distant year in the past. If we use a new starting point which is s units smaller, then the quantity which was originally represented by the number x is now represented by the new value $x + s$. The transformation $x \to x + s$ is usually called a *shift*.

Together, scaling and shifts form *linear transformations* $x \to a \cdot x + b$.

Basic symmetries lead to the following natural requirement: that the physical formulas should not depend on the choice of a measuring unit or of a starting point. In mathematical terms, this means that the physical formulas be invariant under linear transformations.

Examples of Using Symmetries. Let us give two examples of the use of symmetries in physics:

- a simpler example in which we are able to perform all the computations – but the result is not that physically interesting, and
- a more complex example which is physically interesting – but in which we skip all the computations and proofs.

A more detailed description of the use of symmetries in physics can be found, e.g., in [17, 35, 38].

First Example: Pendulum. As the first simple example, let us consider the problem of finding how the period T of a pendulum depends on its length L and on the free fall acceleration g on the corresponding planet. We denote the desired dependence by $T = f(L,g)$. This dependence was originally found by using Newton's equations. We show that (modulo a constant) the same dependence can be obtained without using any differential equations, only by taking the corresponding symmetries into account.

What are the natural symmetries here? To describe a numerical value of the length, we need to select a unit of length. In this problem, there is no fixed length, so it makes sense to assume that the physics does not change if we simply change the unit of length. If we change a unit of length to a one λ times smaller, we get new numerical value $L' = \lambda \cdot L$; e.g., 1.7 m = 170 cm.

Similarly, if we change a unit of time to a one which is μ times smaller, we get a new numerical value for the period $T' = \mu \cdot T$. Under these transformations, the numerical value of the acceleration changes as $g \to g' = \lambda \cdot \mu^{-2} \cdot g$.

Since the physics does not change by simply changing the units, it makes sense to require that the dependence $T = f(L,g)$ also does not change if we simply change the units, i.e., that $T = f(L,g)$ implies $T' = f(L',g')$. Substituting the above expressions for T', L', and g' into this formula, we conclude that $f(\lambda \cdot L, \lambda \cdot \mu^{-2} \cdot g) = \mu \cdot f(L,g)$. From this formula, we can find the explicit expression for the desired function $f(L,g)$. Indeed, let us select λ and μ for which $\lambda \cdot L = 1$ and $\lambda \cdot \mu^{-2} \cdot g = 1$. Thus, we take $\lambda = L^{-1}$ and $\mu = \sqrt{\lambda \cdot g} = \sqrt{g/L}$. For these values λ and μ, the above formula takes the form $f(1,1) = \mu \cdot f(L,g) = \sqrt{g/L} \cdot f(L,g)$. Thus, $f(L,g) = \text{const} \cdot \sqrt{L/g}$ (for the constant $f(1,1)$).

What Is the Advantage of Using Symmetries? The above formula for the period of the pendulum is exactly the same formula that we obtain from Newton's equations.

At first glance, this makes the above derivation of the pendulum formula somewhat useless: we did not invent any new mathematics, the above mathematics is very simple, and we did not get with any new physical conclusion – the formula for the period of the pendulum is well known. Yes, we got a slightly simpler derivation, but once a result is proven, getting a new shorter proof is not very interesting. So what is new in this derivation?

What is new is that we derived the above without using any specific differential equations – we used only the fact that these equations do not have any fixed unit of length or fixed unit of time. Thus, the same formula is true not only for Newton's equations, but also for *any* alternative theory – as long as this alternative theory has the same symmetries.

Another subtle consequence of our result is related to the fact that physical theories need to be experimentally confirmed. Usually, when a formula obtained from a theory turned out to be experimentally true, this is a strong argument for confirming that the original theory is true. One may similarly think that if the pendulum formula is experimentally confirmed, this is a strong argument for confirming that Newton's mechanics is true. However, the fact that we do not need the whole theory to derive the pendulum formula – we only need symmetries – shows that:

- if we have an experimental confirmation of the pendulum formula,
- this does not necessarily mean that we have confirmed Newton's equations – all we confirmed are the symmetries.

Second Example: Shapes of Celestial Objects. Another example where symmetries are helpful is the description of observed geometric shapes of celestial bodies. Many galaxies have the shape of planar logarithmic spirals; other clusters, galaxies, galaxy clusters have the shapes of the cones, conic spirals, cylindrical spirals, straight lines, spheres, etc. For several centuries, physicists have been interested in explaining these shapes. For example, there exist several dozen different physical theories that explain the observed logarithmic spiral shape of many galaxies. These theories differ in their physics, in the resulting differential equations, but they all lead to exactly the same shape – of the logarithmic spiral.

It turns out that there is a good explanation for this phenomenon – all observed shapes can be deduced from the corresponding symmetries; see, e.g., [39, 40, 41, 86]. Here, possible symmetries include shifts, rotations, and "scaling" (dilation) $x_i \to \lambda \cdot x_i$.

The fact that the shapes can be derived from symmetry shows that the observation of these shapes does not confirm one of the alternative theories – it only confirms that all these theories are invariant under shift, rotation, and dilation. This derivation also shows that even if the actual physical explanation for the shape of the galaxies turns out to be different from any of the current competing theories, we should not expect any new shapes – as long as we assume that the physics is invariant with respect to the above basic geometric symmetries.

Symmetries and Similarities Are Actively Used in Physics. The fact that we could derive the pendulum formula so easily shows that maybe in more complex situations, when solving the corresponding differential equation is not as easy, we would still be able to find an explicit solution by using appropriate symmetries. This is indeed the case in many complex problems; see, e.g., [17, 35, 38].

Moreover, in many situations, even equations themselves can be derived from the symmetries. This is true for most equations of fundamental physics: Maxwell's equations of electrodynamics, Einstein's General Relativity equations for describing the gravitation field, Schrödinger's equations of quantum mechanics, etc.; see, e.g., [42, 43].

As a result, in modern physics, often, new theories are formulated not in terms of differential equations, but in term of symmetries. This started with quarks whose theory was first introduced by M. Gell-Mann by postulating appropriate symmetries.

From Linear to Nonlinear Symmetries. Previously, we only considered *linear* symmetries. i.e., transformations which are described by linear functions. Sometimes, however, a system also has *nonlinear* symmetries.

To find such non-linear symmetries, let us recall the general way symmetries are described in mathematics. In general, if a system is invariant under the transformations f and g, then:

- it is invariant under their composition $f \circ g$, and
- it is invariant under the inverse transformation f^{-1}.

In mathematical terms, this means that the corresponding transformations form a *group*. A special branch of mathematics called *group theory* studies such transformation groups and studies properties which are invariant under these transformations.

In practice, at any given moment of time, we can only store and describe finitely many parameters. Thus, it is reasonable to restrict ourselves to *finite-dimensional* groups.

One of the first researcher to explore this idea was Norbert Wiener, the father of cybernetics. He formulated a question [154]: describe all finite-dimensional groups that contain all linear transformations. For transformations from real numbers to real numbers, the answer to this question is known (see, e.g., [119]): all elements of this group are fractionally-linear functions $x \rightarrow \dfrac{a \cdot x + b}{c \cdot x + d}$.

Symmetries and Similarities Are Also Useful beyond Physics. Nonlinear symmetries can be used to explain many semi-empirical computer-related formulas in neural networks, fuzzy logic, etc.; see, e.g., [119].

Independence as Another Example of Symmetry and Similarity. In many real-life situations, we encounter complex systems that consist of a large number of smaller objects (or subsystems) – e.g., molecules that consist of a large number of atoms. In general, the more subsystems we have, the more complex the corresponding models, the more difficult their algorithmic analysis – because in general, we need to take into account possible interactions between different subsystems.

In practice, however, this analysis is often made simpler if we know that some of these subsystems are reasonably *independent* – in the sense that changes in some subsystems do not affect other subsystems. This independence is also a form of symmetry – the transformations performed on one of the subsystems does not change the state of the other.

Discrete Symmetries and Related Similarities. When we described geometric symmetries, we mentioned that while in some cases, we have a continuous family of symmetries, in other cases, we have a discrete set of transformations under which the object is invariant; see, e.g., [17, 35, 38]. This is true not only for geometric symmetries, it is true for general symmetries as well.

For example, in electromagnetism, the formulas do not change if we simply replace all positive charges with negative ones and vice versa: particles with opposite charges will continue to attract each other and particles with the same charges will continue to repel each other with exactly the same force.

Similarly, when we use logic to analyze the truth value of different properties and their logical combinations, it usually does not matter whether we take, as a basis, a certain property P (such as "small") or its negation $P' \overset{\text{def}}{=} \neg P$ (such as "large"): we can easily transform the corresponding formulas into one another.

Symmetries and Optimization. In many cases, as we have mentioned, it is natural to require that the corresponding model be invariant with respect to the corresponding transformations. In many such cases, this invariance (= symmetry) requirement enables us to determine the corresponding model.

In some practical problems, however, there is no good reason to believe that the corresponding model is invariant with respect to the corresponding transformations. In this case, since we have no reason to restrict ourselves to a small class of possible models, we have many possible models that we can use. Out of all possible models, it is necessary to select the one which is, in some reasonable sense, the best – e.g., the most accurate in describing the real-life phenomena, or the one which is the fastest to compute, etc.

What does the "best" mean? When we say "the best", we mean that on the set of all appropriate models, there is a relation $\succeq$ describing which model is better or equal in quality. This relation must be transitive (if A is better than B, and B is better than C, then A is better than C). This relation is not necessarily asymmetric, because we can have two models of the same quality. However, we would like to require that this relation be *final* in the sense that it should define a unique *best* model A_{opt}, i.e., the unique model for which $\forall B \, (A_{opt} \succeq B)$. Indeed:

- If none of the models is the best, then this criterion is of no use, so there should be *at least one* optimal model.
- If *several* different models are equally best, then we can use this ambiguity to optimize something else: e.g., if we have two models with the same approximating quality, then we choose the one which is easier to compute. As a result, the original criterion was not final: we get a new criterion ($A \succeq_{new} B$ if either A gives a better approximation, or if $A \sim_{old} B$ and A is easier to compute), for which the class of optimal models is narrower. We can repeat this procedure until we get a final criterion for which there is only one optimal model.

It is also reasonable to require that the relation $A \succeq B$ should be invariant relative to natural transformations.

At fist glance, these requirements sound reasonable but somewhat weak. One can show, however, that they are often sufficient to actually find the optimal model – because optimality with respect to an invariant optimality criterion actually leads to invariance:

Definition 1.1. *Let $\mathscr{A}$ be a set, and let G be a group of transformations defined on the set $\mathscr{A}$.*

- *By an* optimality criterion, *we mean a* pre-ordering *(i.e., a transitive reflexive relation) $\preceq$ on the set $\mathscr{A}$.*
- *An optimality criterion is called G-invariant if for all $g \in G$, and for all $A, B \in \mathscr{A}$, $A \preceq B$ implies $g(A) \preceq g(B)$.*
- *An optimality criterion is called* final *if there exists one and only one element $A \in \mathscr{A}$ that is preferable to all the others, i.e., for which $B \preceq A$ for all $B \neq A$.*

Proposition 1.1. [119] *Let $\preceq$ be a G-invariant and final optimality criterion on the class $\mathscr{A}$. Then, the optimal model A_{opt} is G-invariant.*

Proof. Let us prove that the model A_{opt} is indeed G-invariant, i.e., that $g(A_{opt}) = A_{opt}$ for every transformation $g \in G$. Indeed, let $g \in G$. From the optimality of A_{opt}, we conclude that for every $B \in \mathscr{A}$, $g^{-1}(B) \preceq A_{opt}$. From the G-invariance of the optimality criterion, we can now conclude that $B \preceq g(A_{\text{opt}})$. This is true for all $B \in \mathscr{A}$ and therefore, the family $g(A_{\text{opt}})$ is optimal. But since the criterion is final, there is only one optimal family; hence, $g(A_{\text{opt}}) = A_{\text{opt}}$. So, A_{opt} is indeed invariant. The proposition is proven.

Approximate Symmetries and Related Similarities. In many physical situations, we do not have *exact* symmetries, we only have *approximate* symmetries. For example, while a shape of a spiral galaxy can be reasonably well described by a logarithmic spiral, this description is only approximate; the actual shape is slightly different. Actually, most symmetries are approximate (see, e.g., [38]); in some cases, the approximation is so good that we can ignore this approximate character and consider the object to be fully symmetric, while in other cases, we have to take asymmetry into account to get an accurate description of the corresponding phenomena.

What We Do in This Book. Because symmetries and similarities are ubiquitous and useful, in this book, we show how symmetries and similarities can be used in all classes of algorithmic problems of sciences and engineering: from analysis to prediction to control. Specifically, in Chapter 2, we show how an approach based on symmetry and similarity can be used in the analysis of real-life systems; in Chapter 3, we show how this approach can be used in the algorithmics of prediction; and in Chapter 4, we show how the the approach based on symmetry and similarity can be used in the algorithmics of control. Possible ideas for future work are listed in Chapter 5.

Using Symmetries and Similarities in the Analysis of Real-Life Systems: An Overview. Most applications of computing deal with real-life systems: we need to predict the behavior of such systems, we need to find the best way to change their behavior, etc. In all these tasks, we first need to *describe* the system's behavior in precise terms – i.e., in terms understandable to a computer, and then use the resulting description to analyze these systems. In this chapter, following the general ideas of Chapter 1, we show that symmetries can help with such description and analysis.

Many real-life systems consist of several interacting subsystems: a galaxy consists of stars, a solid body consists of molecules, etc. To adequately describe and analyze such systems, we need to describe and analyze the corresponding subsystems and their interaction. Thus, in order to describe and analyze generic real-life systems, we need to first be able to describe basic fundamental systems such as molecules, atoms, elementary particles, etc. Because of this fact, in the present chapter, we concentrate on such fundamental systems.

We start with the cases where we can use the most natural symmetries – continuous families of geometric symmetries such as rotations, shifts, etc. A shape of the molecule is formed by its atoms. In comparison to a molecule, an atom is practically a point. A set consisting of a few atoms is, from the geometric viewpoint, a set of a few points; this few-points shape may have a few symmetries, but, with the exception of linear molecules like H_2, it cannot have a continuous family of symmetries. For a molecule to have a geometric shape allowing a continuous family of symmetries, it needs to contain a large number of atoms. Such molecules are typical in biosciences. Because of this, in Section 2.2, we show how, for biomolecules, the corresponding symmetries and similarities naturally explain the observed shapes.

Non-trivial smaller molecules, as we have mentioned, can only have discrete symmetries. In Section 2.3, we show how the approach based on symmetry and similarity can help in describing such molecules.

Finally, when we get to the level of elementary particles and quantum effects describing their interactions, we, in general, no longer have geometric symmetries. Instead, we have a reasonable symmetry-related physical idea of *independence* – that changes in some subsystems do not affect other subsystems. In Section 2.4, we show that this symmetry-related idea leads to a formal justification of quantum theory (in its Feynman integral formulation).

Overall, we show that symmetries can help with description and analysis of fundamental physical systems.

Using Symmetries and Similarities to Help with Prediction of Real-Life Systems. As we have mentioned earlier, one of the main objectives of science is to predict future events. From this viewpoint, the first question that we need to ask is: *is it possible* to predict? In many cases, predictions are possible, but in many other practical situations, what we observe is a *random* (un-predictable) sequence. The question of how we can check predictability – i.e., check whether the given sequence is random – is analyzed in Section 3.2. In this analysis, we use symmetries – namely, we use scaling symmetries.

In situations where prediction is, in principle, possible, the next questions is: *how* can we predict? In cases where we know the corresponding equations, we can use these equations for prediction. In many practical situations, however, we do not know the equations. In such situations, we need to use general prediction and extrapolation tools, e.g., neural networks. In Section 3.3, we show how discrete symmetries and related similarities can help improve the efficiency of neural networks.

Once the prediction is made, the next question is *how accurate* is this prediction? In Section 3.4, we show how scaling symmetries can help in quantifying the uncertainty of the corresponding model; in Section 3.5, we use similar symmetries to find an optimal way of processing the corresponding uncertainty, and in Section 3.6, on the example of a geophysical application, we estimate the accuracy of *spatially locating* the corresponding measurement results.

From the theoretical viewpoint, the most important question is to generate a prediction, no matter how long it takes to perform the corresponding computations. In practice, however, we often need to have the prediction results by a certain time; in this case, it is important to be able to perform the corresponding computations efficiently, so that we have the results by a given deadline. The theoretical possibility of such efficient computations is analyzed in Section 3.7.

Overall, we show that symmetries and similarities can help with all the algorithmic aspects of prediction.

Using Symmetries and Similarities in Control. In Chapter 3, we concentrate on the problem of predicting the future events. Once we are able to predict future events, a natural next step is to *influence* these events, i.e., to *control* the corresponding system. In this step, we should select a control that leads to the best possible result.

Control problems can be roughly classified into two classes. In some problems of this type, we know the exact equations and we know the objective function that describes what the users want. In such problems, the selection of the best possible control is a mathematically well-defined optimization problem. Problems of this type have been solved for centuries, and there are many efficient algorithms that solve these types of problems in many practical situations.

However, there are situations in which the problems of finding the best control are much more challenging, because we only know the system with a huge uncertainty. Because of the uncertainty, it is difficult to formulate the corresponding problem as a precise optimization problem. Instead, we use intelligent techniques that use the knowledge and experience of human experts in solving such problems. Such intelligent techniques are reasonably new: they have been in use for only a few decades. Details of many of such techniques have been determine purely *empirically*, and as a result, they often lead to results which are far from optimal. To improve the results of applying these techniques, it is therefore imperative to perform a *theoretical* analysis of the corresponding problems. In this chapter, in good accordance with the general ideas from Chapter 1, we show that symmetry-based techniques can be very useful in this theoretical analysis.

We illustrate this usefulness on all level of the problem of selecting the best control. First, on a strategic level, we need to select the best class of strategies. In Section 4.2, we use logical symmetries – the symmetry between true and false values – to find the best class of strategies for an important class of intelligent controls – fuzzy control.

Once a class of strategies is selected, we need to select the best strategy within a given class. We analyze this problem in Sections 4.3 and 4.4: in Section 4.3, we use approximate symmetries to find the best operations for implementing fuzzy control, and in Section 4.4, again in good accordance with Chapter 1, that the optimal selection of operations leads to a symmetry-based solution.

Finally, when we have several strategies coming from different aspects of the problem, we need to combine these strategies into a single strategy that takes all the aspects into account. In Section 4.5, we again use logical symmetries – this time to find the best way of combining the resulting fuzzy decisions.

Overall, we show that symmetries and similarities can help with all the algorithmic aspects of control.

Chapter 2
Algorithmic Aspects of Real-Life Systems Analysis: Approach Based on Symmetry and Similarity

2.1 Describing and Analyzing Real-Life Systems

Most applications of computing deal with real-life systems: we need to predict the behavior of such systems, we need to find the best way to change their behavior, etc. In all these tasks, we first need to *describe* the system's behavior in precise terms – i.e., in terms understandable to a computer, and then use the resulting description to analyze these systems. In this chapter, following the general ideas of Chapter 1, we show that symmetries and similarities can help with such description and analysis.

Many real-life systems consist of several interacting subsystems: a galaxy consists of stars, a solid body consists of molecules, etc. To adequately describe and analyze such systems, we need to describe and analyze the corresponding subsystems and their interaction. Thus, in order to describe and analyze generic real-life systems, we need to first be able to describe basic fundamental systems such as molecules, atoms, elementary particles, etc. Because of this fact, in the present chapter, we concentrate on such fundamental systems.

We start with the cases where we can use the most natural symmetries – continuous families of geometric symmetries such as rotations, shifts, etc. A shape of the molecule is formed by its atoms. In comparison to a molecule, an atom is practically a point. A set consisting of a few atoms is, from the geometric viewpoint, a set of a few points; this few-points shape may have a few symmetries, but with the exception of linear molecules like H_2, it cannot have a continuous family of symmetries. For a molecule to have a geometric shape allowing a continuous family of symmetries, it needs to contain a large number of atoms. Such molecules are typical in biosciences. Because of this, in Section 2.2, we show how, for biomolecules, the corresponding symmetries naturally explain the observed shapes.

Non-trivial smaller molecules, as we have mentioned, can only have discrete symmetries. In Section 2.3, we show how the symmetries approach can help in describing such molecules.

Finally, when we get to the level of elementary particles and quantum effects describing their interactions, we, in general, no longer have geometric symmetries.

© Springer-Verlag Berlin Heidelberg 2015
J. Nava and V. Kreinovich, *Algorithmic Aspects of Analysis, Prediction, and Control in Science and Engineering*, Studies in Systems, Decision and Control 14, DOI: 10.1007/978-3-662-44955-4_2

Instead, we have a reasonable symmetry-related physical idea of *independence* – that changes in some subsystems do not affect other subsystems. In Section 2.4, we show that this symmetry-related idea leads to a formal justification of quantum theory (in its Feynman integral formulation). Feynman integral

Overall, we show that symmetries and similarities can help with description and analysis of fundamental physical systems.

2.2 Towards Symmetry- and Similarity-Based Explanation of (Approximate) Shapes of Alpha-Helices and Beta-Sheets (and Beta-Barrels) in Protein Structure

Alpha-Helices and Beta-Sheets: Brief Reminder. Proteins are biological polymers that perform most of the life's function. A single chain polymer (protein) is folded in such a way that forms local substructures called secondary structure elements. In order to study the structure and function of proteins it is extremely important to have a good geometrical description of the proteins structure. There are two important secondary structure elements: alpha-helices and beta-sheets. A part of the protein structure where different fragments of the polypeptide align next to each other in extended conformation forming a *line-like* feature defines a secondary structure called an *alpha-helix*. A part of the protein structure where different fragments of the polypeptide align next to each other in extended conformation forming a *surface-like* feature defines a secondary structure called a *beta pleated sheet*, or, for short, a *beta-sheet*; see, e.g., [18, 85].

Shapes of Alpha-Helices and Beta-Sheets: First Approximation. The actual shapes of the alpha-helices and beta-sheets can be complicated. In the first approximation, alpha-helices are usually approximated by *cylindrical spirals* (also known as *circular helices* or (cylindrical) *coils*), i.e., curves which, in an appropriate coordinate system, have the form $x = a \cdot \cos(\omega \cdot t)$, $y = a \cdot \sin(\omega \cdot t)$, and $c = b \cdot t$. Similarly, in the first approximation, beta-sheets are usually approximated as *planes*. These are the shapes that we explain in this section.

What We Do in This Section: Our Main Result. In this section, following the ideas of a renowned mathematician M. Gromov [52], we use symmetries and similarities to show that under reasonable assumptions, the empirically observed shapes of cylindrical spirals and planes are indeed the best families of simple approximating sets.

Thus, symmetries and similarities indeed explain why the secondary protein structures consists of alpha-helices and beta-sheets.

Comment. The main result of this section first appeared in [112].

Auxiliary Result: We Also Explain the (Approximate) Shape of Beta-Barrels. The actual shape of an alpha-helix or of a beta-sheet is somewhat different from these first-approximation shapes. In [147], we showed that symmetries can explain some resulting shapes of beta-sheets. In this section, we add, to the basic approximate shapes of a circular helix and a planes, one more shape. This shape is observed when, due to tertiary structure effects, a beta-sheet "folds" on itself, becoming what is called a *beta-barrel*. In the first approximation, beta-barrels are usually approximated by cylinders. So, in this section, we also explain cylinders.

We hope that similar ideas of symmetry and similarity can be used to describe other related shapes. For example, it would be nice to see if a torus shape – when a cylinder folds on itself – can also be explained by ideas of symmetry and similarity.

Possible Future Work: Need for Explaining Shapes of Combinations of Alpha-Helices and Beta-Sheets. A protein usually consists of several alpha-helices and beta-sheets. In some cases, these combinations of basic secondary structure elements have their own interesting shapes: e.g., coils (alpha-helices) sometimes form a *coiled coil*. In this section, we use symmetries and similarities to describe the basic geometric shape of secondary structure elements; we hope that similar ideas of symmetry and similarity can be used to describe the shape of their combinations as well.

Symmetries and Similarities Are Actively Used in Physics. As we have mentioned in Chapter 1, in our use of symmetries and similarities, we have been motivated by the successes of using symmetries in physics; see, e.g., [38]. So, in order to explain our approach, let us first briefly recall how symmetries and similarities are used in physics.

Symmetries in Physics: Main Idea. In physics, we usually know the differential equations that describe the system's dynamics. Once we know the initial conditions, we can then solve these equations and obtain the state of the system at any given moment of time.

It turns out that in many physical situations, there is no need to actually solve the corresponding complex system of differential equations: the same results can be obtained much faster if we take into account that the system has certain *symmetries* (i.e., transformations under which this system does not change). The pendulum and the shapes of celestial bodies examples from Chapter 1 exemplify this principle.

From Physics to Analyzing Shapes of Proteins: Towards the Formulation of the Problem. It is reasonable to assume that the underlying chemical and physical laws do not change under shifts and rotations. Thus, as a group of symmetries, we take the group of all "solid motions", i.e., of all transformations which are composed of shifts and rotations.

Comment. In the classification of shapes of celestial bodies, we also considered scalings (dilations) $x_i \to \lambda \cdot x_i$. Scalings make sense in astrophysics and cosmology. Indeed, in forming celestial shapes of large-scale objects, the main role is played by long-distance interactions like gravity and electromagnetic forces, and the formulas describing these long-distance interactions are scale-invariant. In contrast, on the molecular level – that corresponds to the shapes of the proteins – short-distance interactions are also important, and these interactions are not necessarily scale-invariant.

Thus, in our analysis of protein shapes, we only consider shifts and rotations.

Reasonable Shapes. In chemistry, different shapes are possible. For example, *bounded* shapes like a point, a circle, or a sphere do occur in chemistry, but, due to their boundedness, they usually (approximately) describe the shapes of relatively small molecules like benzenes, fullerenes, etc.

We are interested in relatively large molecules like proteins, so it is reasonable to only consider potentially *unbounded* shapes. Specifically, we want to describe *connected* components of these shapes.

Reasonable Families of Shapes. We do not want to just find one single shape, we want to find *families* of shapes that approximate the actual shapes of proteins. These families contain several parameters, so that by selecting values of all these parameters, we get a shape.

The more parameters we allow, the larger the variety of the resulting shape and therefore, the better the resulting shape can match the observed protein shape.

We are interested in the shapes that describe the secondary structure, i.e., the first (crude) approximation to the actual shape. Because of this, we do not need too many parameters, we should restrict ourselves to families with a few parameters.

We Want to Select the Best Approximating Family. In principle, we can have many different approximating families. Out of all these families, we want to select a one which is the *best* in some reasonable sense – e.g., the one that, on average, provides the most accurate approximation to the actual shape, or the one which is the fastest to compute, etc.

What Does the "Best" Mean? There are many possible criteria for selecting the "best" family. It is not easy even to enumerate all of them – while our objective is to find the families which are the best according to each of these criteria. To overcome this difficulty, we therefore formulate a *general* description of the optimality criteria and provide a general description of all the families which are optimal with respect to different criteria.

As we showed in Chapter 1, a natural formalization of a general optimality criterion is a pre-ordering relation $A \succeq B$ which is final (i.e., has exactly one optimal family) and which is invariant relative to natural geometric symmetries (which are, in this geometric case, shifts and rotations).

At fist glance, these requirements sounds reasonable but somewhat weak. We show, however, that they are sufficient to actually find the optimal families of shapes – and that the resulting optimal shapes are indeed the above-mentioned observed secondary-structure shapes of protein components.

Definitions and the Main Result. Our goal is to choose the best finite-parametric family of sets. To formulate this problem precisely, we must formalize what a finite-parametric family is and what it means for a family to be optimal. In accordance with the above analysis of the problem, both formalizations use natural symmetries. So, we first formulate how symmetries can be defined for families of sets, then what it means for a family of sets to be finite-dimensional, and finally, how to describe an optimality criterion.

Definition 2.2.1. *Let $g : M \to M$ be a 1-1-transformation of a set M, and let A be a family of subsets of M. For each set $X \in A$, we define the result $g(X)$ of applying this transformation g to the set X as $\{g(x) \,|\, x \in X\}$, and we define the result $g(A)$ of applying the transformation g to the family A as the family $\{g(X) \,|\, X \in A\}$.*

In our problem, the set M is the 3-D space $\mathbb{R}^3$. In general, we will assume that M is a *smooth manifold*, i.e., a set with several 1-1 mappings of bounded domains of $\mathbb{R}^n$ into M (*coordinates*) such that on each region which is covered by two different mappings $m, m' : \mathbb{R}^n \to M$, the composition $m^{-1} \circ m : \mathbb{R}^n \to \mathbb{R}^n$ is differentiable; see, e.g., [83].

Definition 2.2.2. *Let M be a smooth manifold. A group G of transformations $M \to M$ is called a* Lie transformation group, *if G is endowed with a structure of a smooth manifold for which the mapping $g, a \to g(a)$ from $G \times M$ to M is smooth.*

In our problem, the group G is the group generated by all shifts and rotations. In the 3-D space, we need three parameters to describe a general shift, and three parameters to describe a general rotation; thus, the group G is 6-dimensional – in the sense that we need six parameters to describe an individual element of this group.

We want to define r-parametric families of sets in such a way that symmetries from G would be computable based on parameters. Formally:

Definition 2.2.3. *Let M and N be smooth manifolds.*

- *By a* multi-valued function $F : M \to N$ *we mean a function that maps each $m \in M$ into a discrete set $F(m) \subseteq N$.*
- *We say that a multi-valued function is* smooth *if for every point $m_0 \in M$ and for every value $f_0 \in F(m)$, there exists an open neighborhood U of m_0 and a smooth function $f : U \to N$ for which $f(m_0) = f_0$ and for every $m \in U$, $f(m) \subseteq F(m)$.*

Definition 2.2.4. *Let G be a Lie transformation group on a smooth manifold M.*

- *We say that a class A of closed subsets of M is* G-*invariant if for every set $X \in A$, and for every transformation $g \in G$, the set $g(X)$ also belongs to the class.*
- *If A is a G-invariant class, then we say that A is a* finitely parametric family of sets *if there exist:*

  - *a (finite-dimensional) smooth manifold V ;*
  - *a mapping s that maps each element $v \in V$ into a set $s(v) \subseteq M$; and*
  - *a smooth multi-valued function $\Pi : G \times V \to V$*

  such that:

  - *the class of all sets $s(v)$ that corresponds to different $v \in V$ coincides with A, and*
  - *for every $v \in V$, for every transformation $g \in G$, and for every $\pi \in \Pi(g, v)$, the set $s(\pi)$ (that corresponds to π) is equal to the result $g(s(v))$ of applying the transformation g to the set $s(v)$ (that corresponds to v).*

- *Let $r > 0$ be an integer. We say that a class of sets B is a r-parametric class of sets if there exists a finite-dimensional family of sets A defined by a triple (V, s, Π) for which B consists of all the sets $s(v)$ with v from some r-dimensional sub-manifold $W \subseteq V$.*

In our example, we consider families of unbounded connected sets.

Definition 2.2.5. *Let $\mathscr{A}$ be a set, and let G be a group of transformations defined on $\mathscr{A}$.*

- *By an* optimality criterion, *we mean a* pre-ordering *(i.e., a transitive reflexive relation) $\preceq$ on the set $\mathscr{A}$.*
- *An optimality criterion is called G-invariant if for all $g \in G$, and for all $A, B \in \mathscr{A}$, $A \preceq B$ implies $g(A) \preceq g(B)$.*
- *An optimality criterion is called* final *if there exists one and only one element $A \in \mathscr{A}$ that is preferable to all the others, i.e., for which $B \preceq A$ for all $B \neq A$.*

Lemma 2.2.1. *Let M be a manifold, let G be a d-dimensional Lie transformation group on M, and let $\preceq$ be a G-invariant and final optimality criterion on the class $\mathscr{A}$ of all r-parametric families of sets from M, $r < d$. Then:*

- *the optimal family A_{opt} is G-invariant; and*
- *each set X from the optimal family is a union of orbits $\{g(x) \mid g \in G_0\}$ of $\geq (d - r)$-dimensional subgroups G_0 of the group G.*

Comment. For readers' convenience, all the proofs are placed at the end of this section.

Theorem 2.2.1. *Let G be a 6-dimensional group generated by all shifts and rotations in the 3-D space $\mathbb{R}^3$, and let $\preceq$ be a G-invariant and final optimality criterion on the class $\mathscr{A}$ of all r-parametric families of unbounded sets from $\mathbb{R}^3$, $r < 6$. Then each set X from the optimal family is a union of cylindrical spirals, planes, and cylinders.*

Conclusion. These shapes correspond exactly to alpha-helices, beta-sheets (and beta-barrels) that we observe in proteins. Thus, the symmetries and similarities indeed explain the observed protein shapes.

Comment. As we have mentioned earlier, spirals, planes, and cylinders are only the first approximation to the actual shape of protein structures. For example, it has been empirically found that for beta-sheets and beta-barrels, general hyperbolic (quadratic) surfaces provide a good second approximation; see, e.g., [123]. It is worth mentioning that the empirical fact that quadratic models provide the best second approximation can also be theoretical explained by using symmetries and similarities [147].

Proof of Lemma 2.2.1. Since the criterion $\preceq$ is final, there exists one and only one optimal family of sets. Let us denote this family by A_{opt}.

1°. Let us first show that this family A_{opt} is indeed G-invariant, i.e., that $g(A_{\text{opt}}) = A_{\text{opt}}$ for every transformation $g \in G$.

Indeed, let $g \in G$. From the optimality of A_{opt}, we conclude that for every $B \in \mathscr{A}$, $g^{-1}(B) \preceq A_{\text{opt}}$. From the G-invariance of the optimality criterion, we can now conclude that $B \preceq g(A_{\text{opt}})$. This is true for all $B \in \mathscr{A}$ and therefore, the family $g(A_{\text{opt}})$ is optimal. But since the criterion is final, there is only one optimal family; hence, $g(A_{\text{opt}}) = A_{\text{opt}}$. So, A_{opt} is indeed invariant.

2°. Let us now show an arbitrary set X_0 from the optimal family A_{opt} consists of orbits of $\geq (d - r)$-dimensional subgroups of the group G.

Indeed, the fact that A_{opt} is G-invariant means, in particular, that for every $g \in G$, the set $g(X_0)$ also belongs to A_{opt}. Thus, we have a (smooth) mapping $g \to g(X_0)$ from the d-dimensional manifold G into the $\leq r$-dimensional set $G(X_0) = \{g(X_0) \mid g \in G\} \subseteq A_{\text{opt}}$. In the following, we denote this mapping by g_0.

Since $r < d$, this mapping cannot be 1-1, i.e., for some sets $X = g'(X_0) \in G(X_0)$, the pre-image $g_0^{-1}(X) = \{g \mid g(X_0) = g'(X_0)\}$ consists of one than one point. By definition of $g(X)$, we can conclude that $g(X_0) = g'(X_0)$ iff $(g')^{-1}g(X_0) = X_0$. Thus, this pre-image is equal to $\{g \mid (g')^{-1}g(X_0) = X_0\}$. If we denote $(g')^{-1}g$ by $\tilde{g}$, we conclude that $g = g'\tilde{g}$ and that the pre-image $g_0^{-1}(X) = g_0^{-1}(g'(X_0))$ is equal to $\{g'\tilde{g} \mid \tilde{g}(X_0) = X_0\}$, i.e., to the result of applying g' to $\{\tilde{g} \mid \tilde{g}(X_0) = X_0\} = g_0^{-1}(X_0)$. Thus, each pre-image $(g_0^{-1}(X) = g_0^{-1}(g'(X_0)))$ can be obtained from one of these pre-images (namely, from $g_0^{-1}(X_0)$) by a smooth invertible transformation g'. Thus, all pre-images have the same dimension D.

We thus have a *stratification* (fiber bundle) of a d-dimensional manifold G into D-dimensional strata, with the dimension D_f of the factor-space being $\leq r$. Thus, $d = D + D_f$, and from $D_f \leq r$, we conclude that $D = d - D_f \geq n - r$.

So, for every set $X_0 \in A_{\text{opt}}$, we have a $D \geq (n - r)$-dimensional subset $G_0 \subseteq G$ that leaves X_0 invariant (i.e., for which $g(X_0) = X_0$ for all $g \in G_0$). It is easy to check that if $g, g' \in G_0$, then $gg' \in G_0$ and $g^{-1} \in G_0$, i.e., that G_0 is a *subgroup* of the group G. From the definition of G_0 as $\{g \mid g(X_0) = X_0\}$ and the fact that $g(X_0)$ is defined by a smooth transformation, we conclude that G_0 is a smooth sub-manifold of G, i.e., a $\geq (n - r)$-dimensional subgroup of G.

To complete our proof, we must show that the set X_0 is a union of orbits of the group G_0. Indeed, the fact that $g(X_0) = X_0$ means that for every $x \in X_0$, and for every $g \in G_0$, the element $g(x)$ also belongs to X_0. Thus, for every element x of the set X_0, its entire orbit $\{g(x) \mid g \in G_0\}$ is contained in X_0. Thus, X_0 is indeed the union of orbits of G_0. The lemma is proven.

Proof of Theorem 2.2.1. In our case, the natural group of symmetries G is generated by shifts and rotations. So, to apply the above lemma to the geometry of protein structures, we must describe all orbits of subgroups of this groups G.

Since we are interested in connected components, we should consider only connected *continuous* subgroups $G_0 \subseteq G$, since such subgroups explain connected shapes.

Let us start with 1-D orbits. A 1-D orbit is an orbit of a 1-D subgroup. This subgroup is uniquely determined by its "infinitesimal" element, i.e., by the corresponding element of the Lie algebra of the group G. This Lie algebra is easy to describe. For each of its elements, the corresponding differential equation (that describes the orbit) is reasonably easy to solve.

2-D forms are orbits of $\geq$ 2-D subgroups, so, they can be enumerated by combining two 1-D subgroups.

Comment. An alternative (slightly more geometric) way of describing 1-D orbits is to take into consideration that an orbit, just like any other curve in a 3-D space, is uniquely determined by its curvature $\kappa_1(s)$ and torsion $\kappa_2(s)$, where s is the arc length measured from some fixed point. The fact that this curve is an orbit of a 1-D group means that for every two points x and x' on this curve, there exists a transformation $g \in G$ that maps x into x'. Shifts and rotations do not change κ_i, they may only shift s (to $s + s_0$). This means that the values of κ_i are constant. Taking constant κ_i, we get differential equations, whose solution leads to the desired 1-D orbits.

The resulting description of 0-, 1-, and 2-dimensional orbits of connected subgroups G_a of the group G is as follows:

0: The only 0-dimensional orbit is a *point*.
1: A generic 1-dimensional orbit is a *cylindrical spiral*, which is described (in appropriate coordinates) by the equations $z = k \cdot \phi$, $\rho = R_0$. Its limit cases are:

   – a *circle* ($z = 0$, $\rho = R_0$);
   – a *semi-line* (*ray*);
   – a *straight line*.

2: Possible 2-D orbits include:

   – a *plane*;
   – a *semi-plane*;
   – a *sphere*; and
   – a *circular cylinder*.

Since we are only interested in unbounded shapes, we end up with the following shapes:

• a cylindrical spiral (with a straight line as its limit case);
• a plane (or a part of the plane), and
• a cylinder.

The theorem is proven.

Symmetry-Related Speculations on Possible Physical Origin of the Observed Shapes. We have provided a somewhat mathematical explanation for the observed shapes. Our theorem explains the shapes, but not how a protein acquires these shapes.

A possible (rather speculative) explanation can be obtained along the lines of a similar symmetry-based explanation for the celestial shapes; see [39, 40, 41, 86].

In the beginning, protein generation starts with a uniform medium, in which the distribution is homogeneous and isotropic. In mathematical terms, the initial distribution of matter is invariant w.r.t. arbitrary shifts and rotations.

The equations that describe the physical forces that are behind the corresponding chemical reactions are invariant w.r.t. arbitrary shifts and rotations. In other words, these interactions are *invariant* w.r.t. our group G. The *initial distribution* was *invariant* w.r.t. G; the *evolution equations are* also *invariant*; hence, at first glance, we should get a G-invariant distribution of matter for all moments of time.

In reality, we do not see such a homogeneous distribution – because this highly symmetric distribution is known to be *unstable*. As a result, an arbitrarily small perturbations cause drastic changes in the matter distribution: matter concentrates in some areas, and shapes are formed. In physics, such symmetry violation is called *spontaneous*.

In principle, it is possible to have a perturbation that changes the initial highly symmetric state into a state with no symmetries at all, but statistical physics teaches us that it is much more probable to have a gradual symmetry violation: first, some of the symmetries are violated, while some still remain; then, some other symmetries are violated, etc.

Similarly, a (highly organized) solid body normally goes through a (somewhat organized) liquid phase before it reaches a (completely disorganized) gas phase.

If a certain perturbation concentrates matter, among other points, at some point a, then, due to invariance, for every transformation $g \in G'$, we observe a similar concentration at the point $g(a)$. Therefore, the shape of the resulting concentration contains, with every point a, the entire *orbit* $G'(a) = \{g(a) \mid g \in G'\}$ of the group G'. Hence, the resulting *shape consists of* one or several *orbits* of a group G'. This is exactly the conclusion we came up with before, but now we have a physical explanation for it.

2.3 Properties of Molecules with Variant Ligands: An Approach Based on Symmetry and Similarity

Formulation of the Problem: Extrapolation Is Needed. In many practical situations, molecules can be obtained from a symmetric "template" molecule like benzene C_6H_6 by replacing some of its hydrogen atoms with *ligands* (other atoms or atom groups). There can be many possible replacements of this type. To avoid time-consuming testing of all possible replacements, it is desirable to test some of the replacements and then extrapolate to others – so that only the promising molecules, for which the extrapolated values are desirable, will have to be synthesized and tested.

For this extrapolation, D. J. Klein and co-authors proposed to use a poset extrapolation technique developed by G.-C. Rota from MIT; see, e.g., [139]. They showed that in many practical situations, this technique indeed leads to accurate predictions of many important quantities; see, e.g., [29, 62, 63, 61, 64, 67, 68].

One of the limitations of this approach is that this technique has been originally proposed on a heuristic basis, with no convincing justification of its applicability to chemical (or other) problems. In [113], we showed that for the case where all the ligands are of the same type, the poset technique is actually equivalent to a more familiar (and much more justified) Taylor series extrapolation.

In this section, we show that this equivalence can be extended to the case where we have variant ligands, and that this approach is also equivalent to another well-known and well-studied approach: the Dempster-Shafer approach [155].

Comment. The results of this section first appeared in [99, 105].

Poset Approach to Extrapolation: Reminder. In [139], Gian-Carlo Rota, a renowned mathematician from MIT, considered the situation in which there is a natural partial order relation $\leq$ on some set of objects, and there is a numerical value $v(a)$ associated to each object a from this partially ordered set (poset).

Rota's technique is based on the fact that we can represent an arbitrary dependence $v(a)$ as

$$v(a) = \sum_{b:\,b\leq a} V(b) \tag{2.1}$$

for some values $V(b)$. The possibility to find such values $V(b)$ is easy to understand: the above formula (2.1) is a system of linear equations in which we have as many unknowns $V(b)$ as the number of objects – and as many equations as the number of objects. Thus, we have a system of linear equations with as many equations as there are unknowns. It is known that in general, such a system always has a solution. (In principle, there are degenerate cases where a system of n linear equations with n unknowns does not have a solution, but in [139] it was proven that the poset-related system (2.1) always has a solution.)

In practice, many values $V(b)$ turn out to be negligible and thus, can be safely taken as 0s. If we know which values $V(b_1), \ldots, V(b_m)$ are non-zeros, we can then:

- measure the value $v(a_1), \ldots, v(a_p)$ of the desired quantity v for $p \ll n$ different objects $a_1, \ldots, a_p$;
- use the Least Squares techniques (see, e.g. [143]) to estimate the values $V(b_j)$ from the system

$$v(a_i) = \sum_{j:\,b_j\leq a_i} V(b_j), \; i = 1, \ldots, p; \tag{2.2}$$

- use the resulting estimates $V(b_j)$ to predict all the remaining values $v(a)$ ($a \neq a_1, \ldots, a_m$), as

$$v(a) = \sum_{j:\,b_j\leq a} V(b_j). \tag{2.3}$$

In chemistry, objects are molecules, and a natural relation $a \leq b$ means that the molecule b either coincides with a, or can be obtained from the molecule a if we replace one or several of its H atoms with ligands.

Role of Symmetries. Molecules such as benzene C_6H_6 or cubane C_8H_8 have the property of *symmetry*: for every two atoms, we can find a rotation that moves the first atom into the position of the second one while keeping the molecule configuration intact. For example, for benzene, rotation by $60°$ transforms the first atom into the second one, the second into the third one, etc.

A simple rotation does not change the chemical properties of a molecule – and hence, does not change the values of any numerical property of the substance. Let us start with *monosubstituted* molecules, i.e., molecules in which a single ligand has been substituted. All the monosubstituted molecules can be obtained from each other by rotation. We can therefore conclude that all these molecules have the same values $v(a)$ of all the numerical quantities.

Similarly, when we add two ligands, the value $v(a)$ depends only on the distance between the corresponding locations.

The resulting symmetry of the values $v(a)$ translates into the symmetry of the auxiliary values $V(a)$ as well. Thus, because of the symmetries, some of the unknowns $V(a)$ coincide – i.e., we need to solve systems with fewer unknowns. In short, the (discrete) symmetries of the template molecule help us decrease the number of unknowns and thus, make the corresponding computations easier and faster.

Relation to the Dempster-Shafer Approach. From the purely mathematical viewpoint, formula (2.3) is identical to one of the main formulas of the Dempster-Shafer approach (see, e.g., [155]). Specifically, in this approach,

- in contrast to a probability distribution on a set X where probabilities $p(x) \geq 0$, $\sum_{x \in X} p(x) = 1$, are assigned to different elements $x \in X$ of the set X,
- we have "masses" (in effect, probabilities) $m(A) \geq 0$, $\sum_A m(A) = 1$, assigned to *subsets $A \subseteq X$ of the set X.*

The usual meaning of the values $m(B)$ is, e.g., that we have several experts who have different opinions on which alternatives are possible and which are not. For each expert, B is the set of alternatives that are possible according to this expert, and $m(B)$ is the probability that this expert is correct (estimated, e.g., based on his or her previous performance).

For every set $A \subseteq X$ and for every expert, if the expert's set B of possible alternatives is contained in A ($B \subseteq A$), this means that this expert is sure that all possible alternatives are contained in the set A. Thus, our overall belief $\mathrm{bel}(A)$ that the actual alternative is contained in A can be computed as the sum of the masses corresponding to all such experts, i.e., as

$$\mathrm{bel}(A) = \sum_{B \subseteq A} m(B).$$

This is the exact analog of the above formula, with $v(a)$ instead of belief, $V(b)$ instead of masses, and the subset relation $B \subseteq A$ as the ordering relation $b \leq a$.

Comment. It should be mentioned that in spite of the above similarity, Rota's poset approach is somewhat different from the Dempster-Shafer approach:

- first, in the Dempster-Shafer approach, we require that all the masses are non-negative, while in the poset approach, the corresponding values $V(b)$ can be negative as well;

- second, in the Dempster-Shafer approach, we require that the sum of all the masses is 1, while in the poset approach, the sum of all the values $V(b)$ can be any real number.

Traditional (Continuous) Taylor Series: A Brief Reminder. As promised, we now show that the Gian-Carlo Rota's approach is also equivalent to the Taylor series approach. To describe this equivalence, let us first recall the main ideas behind the traditional Taylor series.

Traditionally, in physical and engineering applications, most parameters $x_1, \ldots, x_n$ (such as coordinates, velocity, etc.) are *continuous* – in the sense that their values can continuously change from one value to another. The dependence $y = f(x_1, \ldots, x_n)$ of a quantity y on the parameters x_i is also usually continuous (with the exception of phase transitions); moreover, this dependence is usually smooth (differentiable). It is known that smooth functions can be usually expanded into Taylor series around some point $\tilde{x} = (\tilde{x}_1, \ldots, \tilde{x}_n)$ (e.g., around the point $\tilde{x} = 0$), i.e., as a sum of constant terms, linear terms, quadratic terms, and terms of higher order.

$$f(x_1, \ldots, x_n) = f(\tilde{x}_1, \ldots, \tilde{x}_n) + \sum_{i=1}^{n} \frac{\partial f}{\partial x_i} \cdot \Delta x_i + \frac{1}{2} \sum_{i=1}^{n} \sum_{i'=1}^{n} \frac{\partial^2 f}{\partial x_i \partial x_{i'}} \cdot \Delta x_i \cdot \Delta x_{i'} + \ldots,$$

where $\Delta x_i \overset{\text{def}}{=} x_i - \tilde{x}_i$.

The values of different order terms in the Taylor expansion usually decrease when the order increases – after all, the Taylor series usually converges, which implies that the terms should tend to 0. So, in practice, we can ignore higher-order terms and consider only the first few terms in the Taylor expansion. (This is, for example, how most elementary functions like $\sin(x)$, $\cos(x)$, $\exp(x)$ are computed inside the computers.).

In the simplest case, it is sufficient to preserve linear terms, i.e. to use the approximation

$$f(x_1, \ldots, x_n) \approx f(\tilde{x}_1, \ldots, \tilde{x}_n) + \sum_{i=1}^{n} \frac{\partial f}{\partial x_i} \cdot \Delta x_i.$$

When the linear approximation is not accurate enough, we can use the quadratic approximation

$$f(x_1, \ldots, x_n) \approx f(\tilde{x}_1, \ldots, \tilde{x}_n) + \sum_{i=1}^{n} \frac{\partial f}{\partial x_i} \cdot \Delta x_i + \frac{1}{2} \cdot \sum_{i=1}^{n} \sum_{i'=1}^{n} \frac{\partial^2 f}{\partial x_i \partial x_{i'}} \cdot \Delta x_i \cdot \Delta x_{i'},$$

etc.

Since we do not know the exact expression for the function $f(x_1, \ldots, x_n)$, we thus do not know the actual values of its derivatives $\dfrac{\partial f}{\partial x_i}$ and $\dfrac{\partial^2 f}{\partial x_i \partial x_{i'}}$. Hence, when we actually use this approximation, all we know is that we approximate a general function by a general linear or quadratic formula

$$f(x_1, \ldots, x_n) \approx c_0 + \sum_{i=1}^{n} c_i \cdot \Delta x_i,$$

$$f(x_1, \ldots, x_n) \approx c_0 + \sum_{i=1}^{n} c_i \cdot \Delta x_i + \sum_{i=1}^{n} \sum_{i'=1}^{n} c_{ii'} \cdot \Delta x_i \cdot \Delta x_{i'}, \tag{2.4}$$

where $c_0 = f(\tilde{x}_1, \ldots, \tilde{x}_n)$, $c_i = \dfrac{\partial f}{\partial x_i}$, and $c_{ii'} = \dfrac{1}{2} \cdot \dfrac{\partial^2 f}{\partial x_i \partial x_{i'}}$.

In the traditional physical and engineering applications, the values of the coefficients c_0, c_i, and (if needed) $c_{ii'}$ can then be determined experimentally. Namely, in several (E) different experiments $e = 1, 2, \ldots, E$, we measure the values $x_i^{(e)}$ of the parameters and the resulting value $y^{(e)}$, and then determine the desired coefficients by applying the Least Squares method to the corresponding approximate equations. In the case of linear dependence, we use approximate equations

$$y^{(e)} \approx c_0 + \sum_{i=1}^{n} c_i \cdot \Delta x_i^{(e)}; \quad e = 1, 2, \ldots, E. \tag{2.5}$$

In the case of quadratic dependence, we use approximate equations

$$y^{(e)} \approx c_0 + \sum_{i=1}^{n} c_i \cdot \Delta x_i^{(e)} + \sum_{i=1}^{n} \sum_{i'=1}^{n} c_{ii'} \cdot \Delta x_i^{(e)} \cdot \Delta x_{i'}^{(e)}. \tag{2.6}$$

From Continuous to Discrete Taylor Series. As we have mentioned in [113], we can extend the Taylor series approach to the discrete case.

In our chemical problem, the discrete case means that for each location, we are only interested in the values of the desired physical quantity in the following situations:

- a situation where there is a ligand at this location, and
- a situation where there is no ligand at this location.

From the macroscopic viewpoint, there are only these two options. However, on the microscopic level, the situation is more complex. Chemical interactions are, in effect, interaction of electrons. A proper description of an electron requires quantum physics; see, e.g., [38].

In classical (pre-quantum) physics, to describe the state of a particle at any given moment of time, it is sufficient to describe its spatial location and momentum. We may not know the exact values of these quantities, but in principle, we can determine them with an arbitrarily high accuracy.

In contrast, in quantum physics, it is impossible to uniquely determine both spatial location and momentum, we can only predict probabilities of different spatial locations (and different momentum values). A quantum description of the state of a particle is a *wave function* $\psi(x)$, a complex-valued function for which, for a small neighborhood of volume ΔV around a point x, the probability to find the electron in this neighborhood is approximately equal to $|\psi(x)|^2 \cdot \Delta V$. In other words, $|\psi(x)|^2$ is the probability density – electronic density in case of electrons.

In principle, electrons can be in many different states, with different electronic density functions $|\psi(x)|^2$. In chemistry, we usually consider only the stable (lowest energy) states. From this viewpoint, we have one of the two situations:

- a situation in which there is a ligand at this location; in this case, we consider the lowest-energy state of a molecule with a ligand at this location; and
- a situation in which there is no ligand at this location; in this case, we consider the lowest-energy state of a molecule with no ligand at this location.

However, from the physical viewpoint, it also makes sense to consider "excited" (higher-energy) states as well, states with arbitrary (not necessarily lowest-energy) electron density functions. Many such states occur as intermediate states in chemical reactions, when a molecule or a group of molecules continuously moves from the original stable state (before the reaction) to a new stable state (after the reaction).

The general physical laws and dependencies are not limited to the discrete (lowest-energy) situations, they work for other (not so stable) situations as well.

So, while we are interested in the values of the desired physical quantity (such as energy) corresponding to the selected stable situations, in principle, we can consider this dependence for other (not so stable) situations as well. The value of, e.g. energy, depends on the values of the electronic density at different points near the ligand locations, etc. For each possible placement of a ligand of type k $(1 \leq k \leq m)$ at a location i $(1 \leq i \leq n)$, let $x_{ik1}, \ldots, x_{ikj}, \ldots, x_{ikN}$ be parameters describing the distribution in the vicinity of this location (e.g., the density at a certain point, the distance to a certain atom, the angle between this atom and the given direction, the angle describing the direction of the spin, etc.). In general, the value of the desired quantity depends on the values of these parameters:

$$y = f(x_{111}, \ldots, x_{11N}, \ldots, x_{nm1}, \ldots, x_{nmN}). \tag{2.7}$$

We are interested in the situations in which, at each location, there is either a ligand, or there is no ligand. For each location i and for each parameter x_{ij}:

- let d_{i0j} denote the value of the j-th parameter in the situation with no ligand at the location i, and
- let d_{ikj} denote the value of the j-th parameter in the situation with a ligand of type k at the location i.

The default situation with which we start is the situation in which there are no ligands at all, i.e. in which $x_{ij} = d_{i0j}$ for all i and j. Other situations of interest are reasonably close to this one. Thus, we can expand the dependence (2.7) in Taylor series in the vicinity of the values d_{i0j}. As a result, we obtain the following expression:

$$y = y_0 + \sum_{i=1}^{n} \sum_{j=1}^{N} y_{ij} \cdot \Delta x_{ij} + \sum_{i=1}^{n} \sum_{j=1}^{N} \sum_{i'=1}^{n} \sum_{j'=1}^{N} y_{ij,i'j'} \cdot \Delta x_{ij} \cdot \Delta x_{i'j'}, \tag{2.8}$$

where $\Delta x_{ij} \overset{\text{def}}{=} x_{ij} - d_{i0j}$, and y_0, y_{ij}, and $y_{ij,i'j'}$ are appropriate coefficients.

These formulas can be applied to all possible situations, in which at each location i, different parameters $x_{i1}, \ldots, x_{iN}$ can change independently. Situations in which we are interested are characterized by describing, for each location, whether there is a ligand or not, and if yes, exactly which ligand. Let ε_{ik} denote the discrete variable that describes the presence of a ligand of type k at the location i:

- when there is no ligand of type k at the location i, we take $\varepsilon_{ik} = 0$, and
- when there is a ligand of type k at the location i, we take $\varepsilon_{ik} = 1$.

By definition, at each location, there can be only one ligand, i.e., if $\varepsilon_{ik} = 1$ for some k, then $\varepsilon_{ik'} = 0$ for all $k' \neq k$.

According to the formula (2.8), the value y of the desired physical quantity depends on the differences Δx_{ij} corresponding to different i and j. Let us describe the values of these differences in terms of the discrete variables ε_{ik}.

- In the absence of a ligand, when $\varepsilon_i = 0$, the value of the quantity x_{ij} is equal to d_{i0j} and thus, the difference Δx_{ij} is equal to

$$\Delta x_{ij} = d_{i0j} - d_{i0j} = 0.$$

- In the presence of a ligand of type k, when $\varepsilon_{ik} = 1$, the value of the quantity x_{ij} is equal to d_{ikj} and thus, the difference $\Delta x_{ij} = d_{ikj} - d_{i0j}$ is equal to

$$\Delta_{ikj} \stackrel{\text{def}}{=} d_{ikj} - d_{i0j}.$$

Taking into account that for each location i, only one value ε_{ik} can be equal to 1, we can combine the above two cases into a single expression

$$\Delta x_{ij} = \sum_{k=1}^{m} \varepsilon_{ik} \cdot \Delta_{ikj}. \tag{2.9}$$

Substituting the expression (2.9) into the expression (2.8), we obtain an expression which is quadratic in ε_{ik}:

$$y = y_0 + \sum_{i=1}^{n} \sum_{k=1}^{m} \sum_{j=1}^{N} y_{ij} \cdot \varepsilon_{ik} \cdot \Delta_{ikj} +$$

$$\sum_{i=1}^{n} \sum_{k=1}^{m} \sum_{j=1}^{N} \sum_{i'=1}^{n} \sum_{k'=1}^{m} \sum_{j'=1}^{N} y_{ij,i'j'} \cdot \varepsilon_{ik} \cdot \varepsilon_{i'k'} \cdot \Delta_{ikj} \cdot \Delta_{i'k'j'}, \tag{2.10}$$

i.e., equivalently,

$$y = y_0 + \sum_{i=1}^{n} \left(\sum_{k=1}^{m} \sum_{j=1}^{N} y_{ij} \cdot \Delta_{ikj} \right) \cdot \varepsilon_{ik} +$$

$$\sum_{i=1}^{n} \sum_{i'=1}^{n} \left(\sum_{j=1}^{N} \sum_{k=1}^{m} \sum_{j'=1}^{N} \sum_{k'=1}^{m} y_{ij,i'j'} \cdot \Delta_{ikj} \cdot \Delta_{i'k'j'} \right) \cdot \varepsilon_{ik} \cdot \varepsilon_{i'k'}.$$

Combining terms proportional to each variable ε_{ik} and to each product $\varepsilon_{ik} \cdot \varepsilon_{i'k'}$, we obtain the expression

$$y = a_0 + \sum_{i=1}^{n} \sum_{k=1}^{m} a_{ik} \cdot \varepsilon_{ik} + \sum_{i=1}^{n} \sum_{k=1}^{m} \sum_{i'=1}^{n} \sum_{k'=1}^{m} a_{ik,i'k'} \cdot \varepsilon_{ik} \cdot \varepsilon_{i'k'}, \qquad (2.11)$$

where

$$a_{ik} = \sum_{j=1}^{N} y_{ij} \cdot \Delta_{ikj}, \qquad (2.12)$$

and

$$a_{ik,i'k'} = \sum_{j=1}^{N} \sum_{j'=1}^{N} y_{ij,i'j'} \cdot \Delta_{ikj} \cdot \Delta_{i'k'j'}. \qquad (2.13)$$

The expression (2.11) is similar to the continuous Taylor expression (2.4), but with the discrete variables $\varepsilon_{ik} \in \{0,1\}$ instead of the continuous variables Δx_i.

Similar "discrete Taylor series" can be derived if we take into account cubic, quartic, etc., terms in the original Taylor expansion of the dependence (2.7).

Discrete Taylor Expansions Can Be Further Simplified. In the following text, we use the fact that the expression (2.11) can be further simplified.

First, we can simplify the terms corresponding to $i = i'$. Indeed, for each discrete variable $\varepsilon_{ik} \in \{0,1\}$, we have $\varepsilon_{ik}^2 = \varepsilon_{ik}$. Thus, the term $a_{ik,ik} \cdot \varepsilon_{ik} \cdot \varepsilon_{ik}$ corresponding to $i = i'$ and $k = k'$ is equal to $a_{ik,ik} \cdot \varepsilon_{ik}$ and can, therefore, be simply added to the corresponding linear term $a_{ik} \cdot \varepsilon_{ik}$.

Similarly, for every location i and for every two ligand types $k \neq k'$, only one of the terms ε_{ik} and $\varepsilon_{ik'}$ can be different from 0. Thus, the product $\varepsilon_{ik} \cdot \varepsilon_{ik'}$ is always equal to 0. Therefore, we can safely assume that the coefficient $a_{ik,ik'}$ at this product is 0.

Thus, we have no terms $a_{ik,i'k'}$ corresponding to $i = i'$ in our formula, we only have terms with $i \neq i'$. For each two pairs ik and $i'k'$, we can combine terms proportional to $\varepsilon_{ik} \cdot \varepsilon_{i'k'}$ and to $\varepsilon_{i'k'} \cdot \varepsilon_{ik}$. As a result, we obtain a simplified expression

$$y = v_0 + \sum_{i=1}^{n} \sum_{k=1}^{m} v_{ik} \cdot \varepsilon_{ik} + \sum_{i<i'}^{} \sum_{k=1}^{m} \sum_{k'=1}^{m} v_{ik,i'k'} \cdot \varepsilon_{ik} \cdot \varepsilon_{i'k'}, \qquad (2.14)$$

where $v_0 = c_0$, $v_{ik} = c_{ik}$, and $v_{ik,i'k'} = c_{ik,i'k'} + c_{i'k',ik}$.

This expression (2.14) – and the corresponding similar cubic and higher order expressions – is what we understand by a discrete Taylor series.

What We Do in the Following Text. As we have mentioned earlier, we show that the poset-related approaches are, in effect, equivalent to the use of a much simpler (and much more familiar) tool of (discrete) Taylor series.

Discrete Taylor Series: Reminder. In many practical situations, we have a physical variable y that depends on the discrete parameters ε_{ik} which take two possible values: 0 and 1, and for which, for every i, at most one value ε_{ik} can be equal to 1.

Then, in the first approximation, the dependence of y on ε_{ik} can be described by the following linear formula

$$y = v_0 + \sum_{i=1}^{n} \sum_{k=1}^{m} v_{ik} \cdot \varepsilon_{ik}. \tag{2.15}$$

In the second approximation, this dependence can be described by the following quadratic formula

$$y = v_0 + \sum_{i=1}^{n} \sum_{k=1}^{m} v_{ik} \cdot \varepsilon_{ik} + \sum_{i<i'} \sum_{k=1}^{m} \sum_{k'=1}^{m} v_{ik,i'k'} \cdot \varepsilon_{ik} \cdot \varepsilon_{i'k'}. \tag{2.16}$$

etc.

Chemical Substances. For chemical substances, we have discrete variables ε_{ik} that describe whether there is a ligand of type k at the i-th location:

- the value $\varepsilon_{ik} = 0$ means that there is no ligand of type k at the i-th location, and
- the value $\varepsilon_{ik} = 1$ means that there is a ligand of type k at the i-th location.

Each chemical substance a from the corresponding family can be characterized by the corresponding tuple

$$(\varepsilon_{11}, \ldots, \varepsilon_{1m}, \ldots, \varepsilon_{n1}, \ldots, \varepsilon_{nm}).$$

Poset-Related Approaches: Reminder. We approximate the actual dependence of the desired quantity y on the substance $a = (\varepsilon_{11}, \ldots, \varepsilon_{nm})$ by a formula

$$v(a) = \sum_{b:b\leq a} V(b), \tag{2.17}$$

where, in the second order approximation, b runs over all substances with at most two ligands.

Poset-Related Approaches Reformulated in Terms of the Discrete Variables. The discrete Taylor series formula (2.16) is formulated in terms of the discrete variables ε_{ik}. Thus, to show the equivalence of these two approaches, let us first describe the poset-related formula (2.17) in terms of these discrete variables.

In chemical terms, the relation $b \leq a$ means that a can be obtained from b by adding some ligands. In other words, the corresponding value ε_{ik} can only increase when we move from the substance b to the substance a. So, if $b = (\varepsilon'_{11}, \ldots, \varepsilon'_{nm})$ and $a = (\varepsilon_{11}, \ldots, \varepsilon_{nm})$, then $b \leq a$ means that for every i and k, we have $\varepsilon'_{ik} \leq \varepsilon_{ik}$.

Thus, the formula (2.17) means that for every substance $a = (\varepsilon_{11}, \ldots, \varepsilon_{nm})$, the substances $b \leq a$ are:

- the original substance $a_0 = (0, \ldots, 0)$;
- substances $a_{ik} \stackrel{\text{def}}{=} (0, \ldots, 0, 1, 0, \ldots, 0)$ with a single ligand of type k at the location i – corresponding to all the places i and types k for which $\varepsilon_{ik} = 1$; and

- substances $a_{ik,i'k'} \overset{\text{def}}{=} (0,\ldots,0,1,0,\ldots,0,1,0,\ldots,0)$ with a ligand of type k at the locations i and a ligand of type k' at a location i' – corresponding to all possible pairs (i,k) and (i',k'), $i < i'$, for which $\varepsilon_{ik} = \varepsilon_{i'k'} = 1$.

Thus, in terms of the discrete variables, the poset formula (2.17) takes the form

$$y = V(a_0) + \sum_{(i,k):\varepsilon_{ik}=1} V(a_{ik}) + \sum_{i<i',k,k':\varepsilon_{ik}=\varepsilon_{i'k'}=1} V(a_{ik,i'k'}). \qquad (2.18)$$

Proof That the Discrete Taylor Series Are Indeed Equivalent to the Poset Formula. The formulas (2.16) and (2.18) are now very similar, so we are ready to prove that they actually coincide.

To show that these formulas are equal, let us take into account that, e.g. the linear part of the sum (2.18) can be represented as

$$\sum_{(i,k):\varepsilon_{ik}=1} V(a_{ik}) = \sum_{(i,k):\varepsilon_{ik}=1} V(a_{ik}) \cdot \varepsilon_{ik}. \qquad (2.19)$$

Indeed, for all the corresponding pairs (i,k), we have $\varepsilon_{ik} = 1$, and multiplying by 1 does not change a number.

This new representation (2.19) allows us to simplify this formula by adding similar terms $V(a_{ik}) \cdot \varepsilon_{ik}$ corresponding to pairs (i,k) for which $\varepsilon_{ik} = 0$. Indeed, when $\varepsilon_{ik} = 0$, then the product $V(a_{ik}) \cdot \varepsilon_{ik}$ is equal to 0, and thus, adding this product does not change the value of the sum. So, in the right-hand side of the formula (2.19), we can safely replace the sum over all pairs (i,k) for which $\varepsilon_{ik} = 1$ by the sum over all pairs (i,k):

$$\sum_{(i,k):\varepsilon_{ik}=1} V(a_{ik}) = \sum_{i=1}^{n} \sum_{k=1}^{m} V(a_{ik}) \cdot \varepsilon_{ik}. \qquad (2.20)$$

Similarly, the quadratic part $\sum_{i<i',k,k':\varepsilon_{ik}=\varepsilon_{i'k'}=1} V(a_{ik,i'k'})$ of the sum (2.18) can be first replaced with the sum

$$\sum_{i<i',k,k':\varepsilon_{ik}=\varepsilon_{i'k'}=1} V(a_{ik,i'k'}) = \sum_{i<i',k,k':\varepsilon_{ik}=\varepsilon_{i'k'}=1} V(a_{ik,i'k'}) \cdot \varepsilon_{ik} \cdot \varepsilon_{i'k'}, \qquad (2.21)$$

and then, by the sum

$$\sum_{i<i',k,k':\varepsilon_{ik}=\varepsilon_{i'k'}=1} V(a_{ik,i'k'}) = \sum_{i<i'} \sum_{k=1}^{m} \sum_{k'=1}^{m} V(a_{ik,i'k'}) \cdot \varepsilon_{ik} \cdot \varepsilon_{i'k'}. \qquad (2.22)$$

Substituting expressions (2.19) and (2.22) into the formula (2.18), we obtain the following expression

$$y = V(a_0) + \sum_{i=1}^{n} V(a_{ik}) \cdot \varepsilon_{ik} + \sum_{i<i'} \sum_{k=1}^{m} \sum_{k'=1}^{m} V(a_{ik,i'k'}) \cdot \varepsilon_{ik} \cdot \varepsilon_{i'k'}. \qquad (2.23)$$

This expression is identical to the discrete Taylor formula (2.16), the only difference is the names of the corresponding parameters:

- the parameter v_0 in the formula (2.16) corresponds to the parameter $V(a_0)$ in the formula (2.23);
- each parameter v_{ik} in the formula (2.16) corresponds to the parameter $V(a_{ik})$ in the formula (2.23); and
- each parameter $v_{ik,i'k'}$ in the formula (2.16) corresponds to the parameter $V(a_{ik,i'k'})$ in the formula (2.23).

The equivalence is proven.

Conclusion. Several practically useful chemical substances can be obtained by adding ligands to different locations of a "template" molecule like benzene C_6H_6 or cubane C_8H_8. here is a large number of such substances, and it is difficult to synthesize all of them and experimentally determine their properties. It is desirable to be able to synthesize and test only a few of these substances and to use appropriate interpolation to predict the properties of others.

It is known that such an interpolation can be obtained by using Rota's ideas related to partially ordered sets. In [113], we have shown that when we only allow one type of ligand, then the exact same interpolation algorithm can be obtained from a more familiar mathematical technique such as Taylor expansion series. In this section, we showed that the similar equivalence holds in the general case, where we have ligands of different type.

2.4 Why Feynman Path Integration: An Explanation Based on Symmetry and Similarity

Need for Quantization. Since the early 1900s, we know that to describe physics properly, we need to take into account quantum effects. Thus, for every non-quantum physical theory describing a certain phenomenon, be it mechanics or electrodynamics or gravitation theory, we must design an appropriate quantum theory.

Traditional Quantization Methods. In quantum mechanics, a state of a physical system is described by a complex-valued *wave function* $\psi(x)$, i.e., a function that associates with each spatial location $x = (x_1, x_2, x_3)$ the value $\psi(x)$ whose meaning is that the square of the absolute value $|\psi(x)|^2$ is a probability density that the measured values of the spatial coordinates will be x. For a system consisting of several particles, a state can similarly be described as a complex-valued function $\psi(x)$ where x is a tuple consisting of all the spatial coordinates of all the particles.

In this formulation, each physical quantity can be described as an *operator* that maps each state into a new complex-valued function. If we know this operator, then we are able to predict how the original state will change when we measure the corresponding physical quantity.

For example, in mechanics, each spatial coordinate x_i is described by an operator that transforms an arbitrary function $\psi(x)$ into a new function $x_i \cdot \psi(x)$, and each component p_i of the momentum vector $\mathbf{p} \overset{\text{def}}{=} m \cdot \mathbf{v}$ is described by the operator that transforms a state $\psi(x)$ into a new state $-i \cdot \hbar \cdot \dfrac{\partial \psi}{\partial x_i}$.

In general, these operators do not commute: e.g., when we first apply the operator x_i and then the operator p_i, then the resulting function will be, in general, different from the result of first applying p_i and then x_i. Since operators describe the process of measuring different physical quantity, this non-commutativity means that the result of measuring a momentum changes if we first measured the spatial coordinates. In other words, when we measure one of these quantities, this measurement process changes the state and thus, changes the result of measuring another quantity. On the example of spatial location and momentum, this phenomenon was first noticed already by W. Heisenberg – a physicist who realized that quantum analogues of physical quantities can be described by operators; this observation that measuring momentum changes the particle's location and vice versa is known as *Heisenberg's uncertainty principle*.

It is known that the equations of quantum mechanics can be written in terms of differential equations describing how the corresponding operators change in time. In the classical (non-quantum) limit, operators turn into the corresponding (commutative) scalar properties, and the corresponding equations turn into the classical Newton's equations. This result has led to the following natural idea of quantizing a theory: replace all the scalars in the classical description of this theory by the corresponding operators.

Limitations of the Traditional Quantization Approach. The problem with the above approach is that due to non-commutativity of the quantum operators, two mathematically equivalent formulations of the classical theory can lead to different (non-equivalent) quantum theories. For example, if in the classical theory, if the expression for the force contains a term proportional to $x_i \cdot v_i$, i.e., proportional both to the spatial coordinates and to the velocities v_i (as, e.g., in the formula for the magnetic force), then we get two different theories depending on whether we replace the above expression with $x_i \cdot \dfrac{p_i}{m}$ or with $\dfrac{p_i}{m} \cdot x_i$.

The corresponding mathematics is non-trivial even in the simplest case, where we only have two unknowns: the location and the momentum of each particle. The situation becomes much more complicated in quantum field theory, where each spatial value $A(x)$ of each field A is a new quantity, with lots of non-commutativity.

Towards Feynman's Approach: The Notion of the Least Action Principle. Starting with Newton's mechanics, the laws of physics have been traditionally described in terms of differential equations, equations that explicitly describe how the rate of change of each quantity depends on the current values of this and other quantities.

For general systems, this is a very good description. However, for *fundamental* physical phenomena – like mechanics, electromagnetic field, etc. – not all differential equations make physical sense. For example, physics usually assumes that fundamental physical quantities such as energy, momentum, and angular momentum, are conserved, so that it is impossible to build a perpetual motion machine which would go forever without source of fuel. However, many differential equations do not satisfy the conservation laws and are, therefore, physically meaningless. It is therefore desirable to restrict ourselves to mathematical models which are consistent with general physical principles such as conservation laws.

It turns out that all known fundamental physical equations can be described in terms of an appropriate optimization principle – and, in general, equations following from a minimization principle lead to conservation laws.

This optimization principle is usually formulated in the following form. Our goal is to find out how the state $\gamma(t)$ of a physical system changes with time t. For example, for a single particle, we want to know how its coordinates (and velocity) change with time. For several particle, we need to know how the coordinates of each particle change with time. For a physical field, we need to know how different field components change with time. Let us assume that we know the state γ at some initial moment of time $\underline{t}$ ($\gamma(\underline{t}) = \underline{\gamma}$), and we know the state $\overline{\gamma}$ at some future moment of time $\overline{t}$ ($\gamma(\overline{t}) = \overline{\gamma}$). In principle, we can have different *paths* (*trajectories*) $\gamma(t)$ which start with the state $\underline{\gamma}$ at the initial moment $\underline{t}$ and end up with the state $\overline{\gamma}$ at the future moment of time $\overline{t}$, i.e., trajectories $\gamma(t)$ for which $\gamma(\underline{t}) = \underline{\gamma}$ and $\gamma(\overline{t}) = \overline{\gamma}$.

For each fundamental physical theory, we can assign, to each trajectory $\gamma(t)$, we can assign a value $S(\gamma)$ such that among all possible trajectories, the actual one is the one for which the value $S(\gamma)$ is the smallest possible. This value $S(\gamma)$ is called *action*, and the principle that action is minimized along the actual trajectory is called the *least action principle*.

For example, between every two spatial points, the light follows the path for which the travel time is the smallest. This fact explains why in a homogeneous medium, light always follows straight paths, and why in the border between two substances, light refracts according to Snell's law.

The motion $x(t)$ of a single particle in a potential field $V(x)$ can also described by the action $S = \int L\,dt$, where $L = \dfrac{1}{2} \cdot \left(\dfrac{dx}{dt}\right)^2 - V(x)$. In general, the action usually takes the form $S = \int L\,dx$, where dx means integration over (4-dimensional) space-time, and the function L is called the *Lagrange function*.

The language of action and Lagrange functions is the standard language of physics: new fundamental theories are very rarely formulated in terms of a differential equation, they are usually formulated in terms of an appropriate Lagrange function – and the corresponding differential equation can then be derived from the resulting minimization principle. Since most classical (non-quantum) physical theories are formulated in terms of the corresponding action $S(\gamma)$, it is reasonable to look for a quantization procedure that would directly transform this action functional into the appropriate quantum theory. Such a procedure was indeed proposed by Richard Feynman, under the name of *path integration*.

Feynman's Path Integration: Main Formulas. In Feynman's approach (see, e.g., [38]), the probability to get from the state $\underline{\gamma}$ to the state $\overline{\gamma}$ is proportional to $\left|\psi\left(\underline{\gamma} \to \overline{\gamma}\right)\right|^2$, where

$$\psi = \sum_{\gamma: \underline{\gamma} \to \overline{\gamma}} \exp\left(i \cdot \frac{S(\gamma)}{\hbar}\right),$$

the sum is taken over all trajectories γ going from $\underline{s}$ to $\overline{s}$, and $\hbar$ is Planck's constant over 2π.

Proportionality means that the transition probability $P\left(\underline{\gamma} \to \overline{\gamma}\right)$ is equal to

$$P\left(\underline{\gamma} \to \overline{\gamma}\right) = C\left(\underline{\gamma}\right) \cdot \left|\psi\left(\underline{\gamma} \to \overline{\gamma}\right)\right|^2,$$

where the proportionality coefficient $C\left(\underline{\gamma}\right)$ should be selected in such a way that the total probability of going from a state $\underline{\gamma}$ to all possible states $\overline{\gamma}$ is equal to 1:

$$\sum_{\overline{\gamma}} P\left(\underline{\gamma} \to \overline{\gamma}\right) = C\left(\underline{\gamma}\right) \cdot \sum_{\overline{\gamma}} \left|\psi\left(\underline{\gamma} \to \overline{\gamma}\right)\right|^2 = 1.$$

Of course, in usual physical situations, there are infinitely many trajectories going from $\underline{\gamma}$ to $\overline{\gamma}$, so instead of the sum, we need to consider an appropriate infinite limit of the sum, i.e., an integral. Thus, in this approach, we integrate over all possible paths; because of this, the approach is called *Feynman path integration*.

Feynman's Path Integration: A Successful Tool. Feynman path integration is not just a foundational idea, it is actually an efficient computing tool. Specifically, as Feynman himself has shown, when we expand Feynman's expression for ψ into the Taylor series, each term in these series can be geometrically represented by a graph. These graphs have physical meaning – they can be interpreted as representing a specific interaction between particles (e.g., that a neutron n^0 gets transformed into a proton p^+, an electron e^-, and an anti-neutrino $\tilde{v}_e$: $n^0 \to p^+ + e^- + \tilde{v}_e$). These graphs are called *Feynman diagrams*. Their physical interpretation helps in computing the corresponding terms. As a result, by adding the values of sufficiently many terms from the Taylor series, we get an efficient technique for computing good approximations to the original value ψ.

This technique has been first used to accurately predict effects in quantum electrodynamics, a theory that has since become instrumental in describing practical quantum-related electromagnetic phenomena such as lasers.

This efficient technique is one of the main reasons why Richard Feynman received a Nobel prize in physics for his research in quantum field theory.

Comment. Feynman's path integration is not only an efficient computational tool: it also helps in explaining methodological questions. For example, in [146], Feynman's path integration is used to explain the common sense meaning of the seemingly counterintuitive Einstein-Podolsky-Rosen experiments – in which, in seeming contradiction to relativity theory, spatially separated particles instantaneously influence each other.

Feynman's Path Integration: Remaining Foundational Question. From the prag-matic viewpoint, Feynman path integral is a great success. However, from the foun-dational viewpoint, we still face an important question: why the above formula?

What We Do in This Section. In this section, we provide a natural explanation for Feynman's path integration formula.

Comment. This explanation first appeared in [102].

This explanation is done on the physical level of rigor, we do not deal here with subtleties of integration as opposed to a simple sum, all we do is justify the general formula – without explaining how to tend to a limit.

Foundations of Feynman Path Integration: Main Ideas and the Resulting Derivation. Let us describe the main ideas that we use in our derivation of path integration.

First Idea: An Alternative Representation of the Original Theory. As a phys-ical theory, we consider a functional S that assigns, to every possible path γ, the corresponding value of the action $S(\gamma)$.

From this viewpoint, a priori, all the paths are equivalent, they only differ by the corresponding values $S(\gamma)$. Thus, what is important for finding out the probability of different transitions is not which value is assigned to different paths, but how many paths are assigned to different values. In other words, what is important is the frequency with which we encounter different values $S(\gamma)$ when selecting a path at random: if among N paths, only one has this value of the action, this frequency is $1/N$, if two, the frequency is $2/N$, etc.

In mathematical terms, this means that we consider the action $S(\gamma)$ as a *random variable* that takes different real values S with different probability.

Because of this analogy, we can use known representations of a random variable to represent the original physical theory. For random variables, there are many dif-ferent representations. One of the most frequently used representations of a random variable α is the *characteristic function* $\xi_\alpha(\omega)$, a function that assigns, to each real number ω, the expected value $E[\cdot]$ of the corresponding exponential expression:

$$\xi(\omega) \stackrel{\text{def}}{=} E[\exp(\mathrm{i} \cdot \omega \cdot \alpha)].$$

In our case, the random variable is $\alpha = S(\gamma)$, where we have N paths γ with equal probability $1/N$. In this case, the corresponding expected value is equal to

$$\xi(\omega) = \frac{1}{N} \cdot \sum_\gamma \exp(\mathrm{i} \cdot S(\gamma) \cdot \omega).$$

Physical comment. One can notice that this expression is similar to Feynman's for-mula for ψ, with $\omega = \dfrac{1}{\hbar}$. This is not yet the desired derivation, because there are many different representations of a random variable, the characteristic function is just one of them. If we use, e.g., a cumulative distribution function or moments, we get completely different formulas. However, this similarity is used in our derivation.

Mathematical comment. For a continuous 1-D random variable, with a probability density function (pdf) $\rho(x)$, the expected value corresponding to the characteristic function takes the form

$$\xi(\omega) \stackrel{\text{def}}{=} E[\exp(\mathrm{i} \cdot \omega \cdot \alpha)] = \int \exp(\mathrm{i} \cdot \omega \cdot x) \cdot \rho(x) \, dx.$$

One can see that, from the mathematical viewpoint, this is exactly the Fourier transform of the pdf (see, e.g., [24]). Thus, once we know the characteristic function $\xi(\omega)$, we can reconstruct the pdf by applying the inverse Fourier transform:

$$\rho(x) \propto \int \exp(\mathrm{i} \cdot \omega \cdot x) \cdot \chi(\omega) \, d\omega$$

(here $\propto$ means "proportional to").

For an n-dimensional random variable with a probability density $\rho(x) = \rho(x_1, \ldots, x_n)$, the characteristic function is similarly equal to the n-dimensional Fourier transform of the pdf and thus, the pdf can be uniquely reconstructed from the characteristic function by applying the inverse n-dimensional Fourier transform.

In general, the probability distribution can be uniquely determined once we know the characteristic function.

Second Idea: Appropriate Behavior for Independent Physical Systems. We want to derive a formula that transforms a functional $S(\gamma)$ into the corresponding transition probabilities. In many cases, the physical system consists of two subsystems. In this case, each state γ of the composite system is a pair $\gamma = (\gamma_1, \gamma_2)$ consisting of a state γ_1 of the first subsystem and the state γ_2 of the second subsystem.

Often, these subsystems are *independent*. Due to this independence, the probability of going from a state $\gamma = (\gamma_1, \gamma_2)$ to a state $\gamma' = (\gamma_1', \gamma_2')$ is equal to the product of the probabilities P_1 and P_2 corresponding to the first and to the second subsystems:

$$P((\gamma_1, \gamma_2) \to (\gamma_1', \gamma_2')) = P_1(\gamma_1 \to \gamma_1') \cdot P_2(\gamma_2 \to \gamma_2').$$

In order to describe how to satisfy this property, let us recall that in physics, independence of two subsystems is usually described by assuming that the corresponding action is equal to the sum of the actions corresponding to two subsystems:

$$S((\gamma_1, \gamma_2)) = S_1(\gamma_1) + S_2(\gamma_2).$$

This relation is usually described in terms of the corresponding Lagrange functions: $L = L_1 + L_2$, which, after integration, leads to the above relation between the actions.

The reason for this sum representation is simple: the actual trajectory is the one that minimizes the total action, i.e., for which the corresponding (variational) derivatives are zeros: $\dfrac{\partial S}{\partial \gamma_1} = 0$ and $\dfrac{\partial S}{\partial \gamma_2} = 0$. When $S((\gamma_1, \gamma_2)) = S_1(\gamma_1) + S_2(\gamma_2)$, we have $\dfrac{\partial S_1(\gamma_2)}{\partial \gamma_1} = \dfrac{\partial S_2(\gamma_2)}{\partial \gamma_1} = 0$ and therefore,

$$\frac{\partial S}{\partial \gamma_1} = \frac{\partial S_1(\gamma_1)}{\partial \gamma_1} + \frac{\partial S_2(\gamma_2)}{\partial \gamma_1} = \frac{\partial S_1(\gamma_1)}{\partial \gamma_1} = 0$$

and

$$\frac{\partial S}{\partial \gamma_2} = \frac{\partial S_1(\gamma_1)}{\partial \gamma_2} + \frac{\partial S_2(\gamma_2)}{\partial \gamma_2} = \frac{\partial S_2(\gamma_2)}{\partial \gamma_2} = 0.$$

Thus, we have two independent systems of equations:

- the system $\dfrac{\partial S_1(\gamma_1)}{\partial \gamma_1} = 0$ that describes the motion of the first subsystem and which does not depend on the second subsystem at all, and
- the system $\dfrac{\partial S_2(\gamma_2)}{\partial \gamma_2} = 0$ that describes the motion of the second subsystem and which does not depend on the first subsystem at all.

Consequences of the Second (Independence) Idea. In probabilistic terms, the fact that $S(\gamma) = S((\gamma_1, \gamma_2)) = S_1(\gamma_1) + S_2(\gamma_2)$ means that the random variable $S(\gamma)$ is the sum of two independent random variables $S_1(\gamma_1)$ and $S_2(\gamma_2)$. To simplify our analysis of this situation, we use the fact that the the characteristic function $\xi_\alpha(\omega)$ of the sum $\alpha = \alpha_1 + \alpha_2$ of two independent random variables α_1 and α_2 is equal to the product of the characteristic functions $\xi_{\alpha_1}(\omega)$ and $\xi_{\alpha_2}(\omega)$ corresponding to these variables: $\xi_\alpha(\omega) = \xi_{\alpha_1}(\omega) \cdot \xi_{\alpha_2}(\omega)$. Indeed, by definition,

$$\xi_\alpha(\omega) = E[\exp(i \cdot \omega \cdot \alpha)] = E[\exp(i \cdot \omega \cdot (\alpha_1 + \alpha_2))] = E[\exp((i \cdot \omega \cdot \alpha_1) + (i \cdot \omega \cdot \alpha_2)].$$

Since $\exp(x + y) = \exp(x) \cdot \exp(y)$, we get

$$\xi_\alpha(\omega) = E[\exp(i \cdot \omega \cdot \alpha)] = E[\exp(i \cdot \omega \cdot \alpha_1) \cdot \exp(i \cdot \omega \cdot \alpha_2)].$$

Due to the fact that the variables α_1 and α_2 are independent, the expected value of the product is equal to the product of expected values:

$$\xi_\alpha(\omega) = E[\exp(i \cdot \omega \cdot \alpha)] = E[\exp(i \cdot \omega \cdot \alpha_1)] \cdot E[\exp(i \cdot \omega \cdot \alpha_2)] = \xi_{\alpha_1}(\omega) \cdot \xi_{\alpha_2}(\omega).$$

Because of this relation, instead of the original dependence of the transition probability p on the functional $S(\gamma)$, we look for the dependence of p on $\xi(\omega)$. We have already argued that all we can use about $S(\gamma)$ is the corresponding probability distribution, and this probability distribution can be uniquely determined by the corresponding characteristic function.

To describe a function, we must describe its values for different inputs. In these terms, we are interested in the dependence p on the variables $\xi(\omega)$ corresponding to different values ω: $p(\xi) = p(\xi(\omega_1), \dots, \xi(\omega_n), \dots)$. The fact that for a combination of two independent systems, the transition probability should be equal to the product of the corresponding probabilities means that

$$p(\xi_1 \cdot \xi_2) = p(\xi_1(\omega_1) \cdot \xi_2(\omega_1), \dots, \xi_1(\omega_n) \cdot \xi_2(\omega_n), \dots) =$$

$$p(\xi_1(\omega_1), \dots, \xi_1(\omega_n), \dots) \cdot p(\xi_2(\omega_1), \dots, \xi_2(\omega_n), \dots).$$

This equation can be further simplified if we move to the log-log scale, i.e., we use $P \stackrel{\text{def}}{=} \ln(p)$ as the new unknown and the values $Z_i = X_i + i \cdot Y_i \stackrel{\text{def}}{=} \ln(\xi(\omega_i))$ as the new variables, i.e., if instead of the original function $p(\xi(\omega_1), \dots, \xi(\omega_n), \dots)$, we consider a new function

$$P(Z_1, \dots, Z_n, \dots) = \ln p(\exp(Z_1), \dots, \exp(Z_n), \dots)$$

Vice versa, once we know the dependence $P(Z_1, \dots, Z_n, \dots)$, we can reconstruct the original function $p(\xi(\omega_1), \dots, \xi(\omega_n), \dots)$ as

$$p(\xi(\omega_1), \dots, \xi(\omega_n), \dots) = \exp(P(\ln(\xi(\omega_1)), \dots, \ln(\xi(\omega_n)), \dots)$$

Since $\ln(x \cdot y) = \ln(x) + \ln(y)$, in terms of P and Z_i, the independence requirement takes the form

$$P(Z_{11} + Z_{21}, \dots, Z_{1n} + Z_{2n}, \dots) = P(Z_{11}, \dots, Z_{1n}, \dots) + P(Z_{21}, \dots, Z_{2n}, \dots),$$

where we denoted $Z_{ij} \stackrel{\text{def}}{=} \ln(\xi_i(\omega_j))$.

Let us further simplify this formula by expressing it in terms of real X_{ij} and imaginary parts Y_{ij} of the real numbers $Z_{ij} = X_{ij} + i \cdot Y_{ij}$. Since

$$\xi_i(\omega_j) = \exp(Z_{ij}) = \exp(X_{ij} + i \cdot Y_{ij}) = \exp(X_{ij}) \cdot \exp(i \cdot Y_{ij}),$$

the real part X_{ij} is equal to the logarithm of the absolute value $|\xi_i(\omega_j)|$ of the complex number $\xi_i(\omega_j)$, and Y_{ij} is equal to the phase of this complex number (recall that every complex number z can be represented as $\rho \cdot \exp(i \cdot \theta)$, where ρ is its absolute value and θ its phase.)

Specifically, we introduce a new function

$$\mathscr{P}(X_1, Y_1, \dots, X_n, Y_n, \dots) \stackrel{\text{def}}{=} P(X_1 + i \cdot Y_1, \dots, X_n + i \cdot Y_n, \dots).$$

In terms of this new function, the above relation takes the form

$$P(X_{11} + X_{21}, Y_{11} + Y_{21}, \dots, X_{1n} + X_{2n}, Y_{1n} + X_{2n}, \dots) =$$

$$P(X_{11}, Y_{11}, \dots, X_{1n}, Y_{1n}, \dots) + P(X_{21}, Y_{21}, \dots, X_{2n}, Y_{2n}, \dots).$$

In other words, for every two tuples $z_1 = (X_{11}, Y_{11}, \dots, X_{1n}, Y_{1n}, \dots)$ and $z_2 = (X_{21}, Y_{21}, \dots, X_{2n}, Y_{2n}, \dots)$, once we define their sum component-wise

$$z_1 + z_2 \stackrel{\text{def}}{=} (X_{11} + X_{21}, Y_{11} + Y_{21}, \dots, X_{1n} + X_{2n}, Y_{1n} + Y_{2n}, \dots),$$

we get

$$\mathscr{P}(z_1 + z_2) = \mathscr{P}(z_1) + \mathscr{P}(z_2).$$

It is known that every continuous (or even measurable) function satisfying this equation is linear, i.e., has the form

$$\mathcal{P}(X_1, Y_1, \ldots, X_n, Y_n, \ldots) = a_1 \cdot X_1 + b_1 \cdot Y_1 + \ldots + a_n \cdot X_n + b_n \cdot Y_n + \ldots$$

(Let us recall that we operate on the physical level of rigor, where we ignore the difference between the finite sum and the limit (infinite) sum such as an integral. In general, since there are infinitely many possible values ω and thus, infinitely many variables $\xi(\omega)$, we need an integral.)

In terms of the function $P(Z_1, \ldots, Z_n, \ldots)$ this formula takes the form

$$P(Z_1, \ldots, Z_n, \ldots) = a_1 \cdot X_1 + b_1 \cdot Y_1 + \ldots + a_n \cdot X_n + b_n \cdot Y_n + \ldots,$$

where $X_i \stackrel{\text{def}}{=} \text{Re}(Z_i)$ and $Y_i \stackrel{\text{def}}{=} \text{Im}(Z_i)$.

Thus, the dependence $p(\xi(\omega_1), \ldots, \xi(\omega_n), \ldots)$ takes the form

$$p(\xi(\omega_1), \ldots, \xi(\omega_n), \ldots) = \exp(P(\ln(\xi(\omega_1)), \ldots, \ln(\xi(\omega_n)), \ldots) =$$

$$\exp(a_1 \cdot X_1 + b_1 \cdot Y_1 + \ldots + a_n \cdot X_n + b_n \cdot Y_n + \ldots).$$

If we represent each complex number $\xi(\omega_i)$ by using its absolute value $\rho(\omega_i) = |\xi(\omega_i)|$ and the phase Y_i, as $\xi(\omega_i) = \rho(\omega_i) \cdot \exp(\mathrm{i} \cdot Y_i)$, then the above formula takes the form

$$p(\xi(\omega_1), \ldots, \xi(\omega_n), \ldots) = \exp(P(\ln(\xi(\omega_1)), \ldots, \ln(\xi(\omega_n)), \ldots) =$$

$$\exp(a_1 \cdot \ln(\rho(\omega_1)) + b_1 \cdot Y_1 + \ldots + a_n \cdot \ln(\rho(\omega_n)) + b_n \cdot Y_n + \ldots).$$

Since $\exp(x+y) = \exp(x) \cdot \exp(y)$, $\exp(a \cdot x) = (\exp(x))^a$, and $\exp(\ln(x)) = x$, we get

$$p(\xi(\omega_1), \ldots, \xi(\omega_n), \ldots) = \exp(P(\ln(\xi(\omega_1)), \ldots, \ln(\xi(\omega_n)), \ldots) =$$

$$\exp(a_1 \cdot \ln(\rho(\omega_1))) \cdot \exp(b_1 \cdot Y_1) \cdot \ldots \exp(a_n \cdot \ln(\rho(\omega_n)) \cdot \exp(b_n \cdot Y_n) \cdot \ldots =$$

$$(\rho(\omega_1))^{a_1} \cdot \exp(b_1 \cdot Y_1 + \ldots \cdot (\rho(\omega_n))^{a_n} \cdot \exp(b_n \cdot Y_n) \cdot \ldots$$

The phase value are determined modulo $2 \cdot \pi$, because $\exp(\mathrm{i} \cdot 2 \cdot \pi) = 1$ and hence,

$$\exp(\mathrm{i} \cdot (Y_i + 2 \cdot \pi)) = \exp(\mathrm{i} \cdot Y_i) \cdot \exp(\mathrm{i} \cdot 2 \cdot \pi) = \exp(\mathrm{i} \cdot Y_i).$$

If we continually change the phase from Y_i to $Y_i + 2 \cdot \pi$, we thus get the same state. The transition probability p_i should therefore not change if we simply add $2 \cdot \pi$ to one of the phases Y_i. When we add $2 \cdot \pi$, the corresponding factor $\exp(b_i \cdot Y_i)$ in the formula for p changes to

$$\exp(b_i \cdot (Y_i + 2 \cdot \pi)) = \exp(b_i \cdot Y_i + b_i \cdot 2 \cdot \pi) = \exp(b_i \cdot Y_i) \cdot \exp(b_i \cdot 2 \cdot \pi).$$

Thus, the only case where the expression for p does not change under this addition is when $\exp(b_i \cdot 2 \cdot \pi) = 1$, i.e., when $b_i \cdot 2 \cdot \pi = 0$ and $b_i = 0$. So, the resulting expression for p takes the form

$$p(\xi(\omega_1), \ldots, \xi(\omega_n), \ldots) = |\xi(\omega_1)|^{a_1} \cdot \ldots \cdot |\xi(\omega_n)|^{a_n} \cdot \ldots$$

We have already mentioned that each value $\xi(\omega_i)$ is equal to the Feynman sum, with an appropriate value of the parameter corresponding to Planck's constant $\hbar_i = 1/\omega_i$. So, we are close to the justification of Feynman's path integration: we just need to explain why in the actual Feynman's formula (the formula which is well justified by the experiments), there is only one factor $|\xi(\omega_i)|$, corresponding to only one value $\hbar_i = 1/\omega_i$. For this, we need a third idea.

Third Idea: Maximal Set of Possible Future States. To explain our third (and final) idea, let us recall one of the main differences between classical and quantum physics:

- in classical (non-quantum) physics, once we know the initial state γ, we can uniquely predict all future states of the system and all future measurement results γ' – of course, modulo the accuracy of the corresponding measurements;
- in quantum physics, even when we have a complete knowledge of the current state γ, we can, at best, predict *probabilities* $p(\gamma \to \gamma')$ of different future measurement results γ'.

In quantum physics, in general, many different measurement results γ' are possible. Of course, it is possible to have $p(\gamma \to \gamma') = 0$: for example, in the traditional quantum mechanics, the wave function may take 0 values at some spatial locations.

From this viewpoint, if we have several candidates for a true quantum theory, we should select a one for which the set of all inaccessible states γ' should be the smallest possible – in the sense that no other candidate theory has a smaller set. Let us look at the consequences of this seemingly reasonable requirement.

As we have mentioned, in general, the transition probability is the product of several expressions $|\xi(\omega_i)|^{a_i}$ corresponding to different values $\hbar_i = 1/\omega_i$. The product is equal to 0 if and only if at least one of these expressions is equal to 0. Thus, the set of all the states γ' for which the product is equal to 0 is equal to the union of the sets corresponding to individual expressions. Thus, the smallest possible set occurs if the whole product consists of only one term, i.e., if

$$p(\xi) = |\xi(\omega)|^a$$

for some constants ω and a.

The only difference between this term and Feynman's formula is that here, we may have an arbitrary value a, while in Feynman's formula, we have $a = 2$.

Fourth Idea: Analyticity and Simplicity. In physics, most dependencies are real-analytical, i.e., expandable in convergent Taylor series. For a complex number $z = x + i \cdot y$, the function $|z|^a = \left(\sqrt{x^2 + y^2}\right)^a = (x^2 + y^2)^{a/2}$ is analytical at $z = 0$ if and only if a is non-negative even number, i.e., $a = 0$, $a = 2$, $a = 4$, ...

The choice $a = 0$ corresponds to a degenerate physical theory $p = $ constant, in which the probability of the transition from a given state γ to every other state γ' is exactly the same. Thus, the simplest non-trivial case is the case of $a = 2$, i.e., the case of Feynman integration.

Physical comments

- The simplicity arguments are, of course, much weaker than the independence arguments given above. Instead of these simplicity arguments, we can simply say that the empirical observation confirms that $p = |\psi|^2$ as in Feynman path integration but not that $p = |\psi|^4$ or $p = |\psi|^6$ etc., as in possible alternative theories.
- The alternatives $a \neq 2$ have a precise mathematical meaning which can be illustrated on the example of a system consisting of a single particle. For this particle, we can talk about the probability of this particle to be at a certain spatial location x. For $a = 2$, this is proportional to $|\psi(x)|^2$, so the fact that the total probability of finding a particle at one of the locations is equivalent to requiring that $\int |\psi(x)|^2 \, dx = 1$, i.e., to the fact that the wave function belongs to the space L^2 of all square integrable functions.

  For $a \neq 2$, the corresponding probability density is proportional to $|\psi(x)|^a$, so we arrive at a similar requirement that $\int |\psi(x)|^a \, dx = 1$. In mathematical terms, this means that the corresponding wave function ψ belongs to the space L^p of all the functions whose p-th power is integrable, for $p = a \neq 2$. It is worth mentioning that the idea of using L^p space has been previously proposed in the foundations of quantum physics; see, e.g., [21, 25, 44, 53].

Chapter 3
Algorithmic Aspects of Prediction: An Approach Based on Symmetry and Similarity

3.1   Algorithmic Aspects of Prediction

As we have mentioned earlier, one of the main objectives of science is to predict future events. From this viewpoint, the first question that we need to ask is: *is it possible* to predict? In many cases, predictions are possible, but in many other practical situations, what we observe is a *random* (un-predictable) sequence. The question of how we can check predictability – i.e., check whether the given sequence is random – is analyzed in Section 3.2. In this analysis, we use symmetries – namely, we use scaling symmetries.

In situations where prediction is, in principle, possible, the next questions is: *how* can we predict? In cases where we know the corresponding equations, we can use these equations for prediction. In many practical situations, however, we do not know the equations. In such situations, we need to use general prediction and extrapolation tools, e.g., neural networks. In Section 3.3, we show how discrete symmetries can help improve the efficiency of neural networks.

Once the prediction is made, the next question is *how accurate* is this prediction? In Section 3.4, we show how scaling symmetries can help in quantifying the uncertainty of the corresponding model; in Section 3.5, we use similar symmetries to find an optimal way of processing the corresponding uncertainty, and in Section 3.6, on the example of a geophysical application, we estimate the accuracy of *spatially locating* the corresponding measurement results.

From the theoretical viewpoint, the most important question is to generate a prediction, no matter how long it takes to perform the corresponding computations. In practice, however, we often need to have the prediction results by a certain time; in this case, it is important to be able to perform the corresponding computations efficiently, so that we have the results by a given deadline. The theoretical possibility of such efficient computations is analyzed in Section 3.7.

Overall, we show that symmetries and similarities can help with all the algorithmic aspects of prediction.

© Springer-Verlag Berlin Heidelberg 2015
J. Nava and V. Kreinovich, *Algorithmic Aspects of Analysis, Prediction, and Control in Science and Engineering*, Studies in Systems, Decision and Control 14, DOI: 10.1007/978-3-662-44955-4_3

3.2   Is Prediction Possible?

Let Us Reformulate This Question in Precise Terms. In many practical situations, predictions are possible, but in many other practical situations, what we observe is a *random* (un-predictable) sequence. How can we tell when a sequence of events is predictable and when it is random? This question was raised in the 1960s by A. N. Kolmogorov, one of the founders of the modern probability theory. Kolmogorov noticed that when a sequence of events allows prediction – i.e., if is a regular sequence like 0101...01 – this means that we have a simple law that enables us to predict future elements of this sequence, i.e., we have a reasonable short program for generating this sequence. On the other hand, if a sequence is truly random, i.e., unpredictable, then it cannot be described by any law, so the only way to generate this sequence is to simply print it. In other words, for a random sequence s, the length of the shortest program that generates this sequence is approximately equal to the length $\text{len}(s)$. Kolmogorov thus defined *Kolmogorov complexity* of a given string s as the shortest length of a program that generates this string; see, e.g., [87].

Once we know the Kolmogorov complexity $K(s)$ of a sequence s, we can tell whether this sequence is random or not: if $K(s) \approx \text{len}(s)$, then the sequence s is random, while if $K(s) \ll \text{len}(s)$ the sequence S is not random.

Kolmogorov Complexity Has Other Applications. Another application of Kolmogorov complexity is that we can check how close are two DNA sequences s and s' by comparing $K(ss')$ with $K(s) + K(s')$:

- if s and s' are unrelated, then the only way to generate ss' is to generate s and then generate s', so $K(ss') \approx K(s) + K(s')$; but
- if s and s' are related, then we have $K(ss') \ll K(s) + K(s')$.

Need for Computable Approximations to Kolmogorov Complexity. The big problem is that the Kolmogorov complexity is, in general, not algorithmically computable [87]. Thus, it is desirable to describe computable approximations to $K(s)$.

Usual Approaches to Approximating Kolmogorov Complexity: Description and Limitations. At present, most algorithms for approximating $K(s)$ use some loss-less compression technique to compress s, and take the length $\widetilde{K}(s)$ of the compression as the desired approximation.

This approximation has limitations. For example, in contrast to $K(s)$, where a small (one-bit) change in x cannot change $K(s)$ much, a small change in s can lead to a drastic change in $\widetilde{K}(s)$.

The General Notion of I-Complexity. To overcome this limitation, V. Becher and P. A. Heiber proposed the following new notion of *I-complexity* [6, 7]. For each position i of the string $s = (s_1 s_2 \ldots s_n)$, we first find the largest $B_s[i]$ of the lengths ℓ of all strings $s_{i-\ell+1} \ldots s_i$ which are substrings of the sequence $s_1 \ldots s_{i-1}$.

Then, we define $I(s) \stackrel{\text{def}}{=} \sum_{i=1}^{n} f(B_s[i])$, for an appropriate decreasing function $f(x)$.

Example. For example, for *aaaab*, the corresponding values of $B_s(i)$ are 01230. Indeed:

- For $i = 1$, the sequence $s_1 \ldots s_{i-1}$ is empty, so $B_s(1) = 0$.
- For $i = 2$, with $s_1 s_2 = aa$, a string $s_2 = a$ is a substring of length 1 of the sequence $s_1 \ldots s_{i-1} = s_1 = a$. So, here, $B_s(2) = 1$.
- For $i = 3$, with $s_1 s_2 s_3 = aaa$, a string $s_2 s_3 = aa$ is a substring of length 2 of the sequence $s_1 \ldots s_{i-1} = s_1 s_2 = aa$. So, here, $B_s(3) = 2$.
- For $i = 4$, with $s_1 s_2 s_3 s_4 = aaaa$, a string $s_2 s_3 s_4 = aaa$ is a substring of length 3 of the sequence $s_1 \ldots s_{i-1} = s_1 s_2 s_3 = aaa$. So, here, $B_s(4) = 3$.
- For $i = 5$, none of the strings $s_{i-\ell+1} \ldots s_i$ ending with $s_i = s_4 = b$ is a substring of the sequence $s_1 \ldots s_{i-1} = s_1 s_2 s_3 s_4 = aaaa$. So, here, $B_s(5) = 0$.

Good Properties of I-Complexity. Thus defined I-complexity has many properties which are similar to the properties of the original Kolmogorov complexity $K(s)$:

- If a string s starts with a substring s', then $I(s) \leq I(s')$.
- We have $I(0s) \approx I(s)$ and $I(1s) \approx I(s)$.
- We have $I(ss') \leq I(s) + I(s')$.
- Most strings have high I-complexity.

On the other hand, in contrast to non-computable Kolmogorov complexity $K(s)$, I-complexity can be computed feasibly: namely, it can be computed in linear time.

Empirical Fact. Which function $f(x)$ should we choose? It turns out that the following *discrete derivative of the logarithm* works the best: $f(x) = \mathrm{dlog}(x+1)$, where $\mathrm{dlog}(x) \stackrel{\mathrm{def}}{=} \log(x+1) - \log(x)$.

Natural Question. How can we explain this empirical fact? In this section, we propose an explanation based on symmetries and similarities. This explanation first appeared in [74, 75].

Discrete Derivatives. Each function $f(n)$ can be represented as the *discrete derivative* $F(n+1) - F(n)$ for an appropriate function $F(n)$: e.g., for $F(n) = \sum_{i=1}^{n-1} f(i)$. In terms of the function $F(n)$, the above question takes the following form: what is the best choice of the function $F(n)$?

From a Discrete Problem to a Continuous Problem. The function $F(x)$ is only defined for integer values x – if we use bits to measure the length of the longest repeated substring. If we use bytes, then x can take rational values, e.g., 1 bit corresponds to 1/8 of a byte, etc. If we use Kilobytes to describe the length, we can use even smaller fractions. In view of this possibility to use different units for measuring length, let us consider the values $F(x)$ for arbitrary real lengths x.

Continuous Quantities: General Observation. In the continuous case, as we have mentioned in Chapter 1, the numerical value of each quantity depends:

- on the choice of the measuring unit and
- on the choice of the starting point.

By changing them, we get a new value $x' = a \cdot x + b$.

Continuous Dependencies: Case of Length x**.** In our case, x is the length of the input. For length x, the starting point 0 is fixed, so we only have re-scaling $x \to \bar{x} = a \cdot x$.

Natural Requirement: The Dependence Should Not Change If We Simply Change the Measuring Unit. When we re-scale x to $\bar{x} = a \cdot x$, the value $y = F(x)$ changes, to $\bar{y} = F(a \cdot x)$. It is reasonable to require that the value $\bar{y}$ represent the same quantity, i.e., that it differs from y by a similar re-scaling: $\bar{y} = F(a \cdot x) = A(a) \cdot F(x) + B(a)$ for appropriate values $A(a)$ and $B(a)$.

Resulting Precise Formulation of the Problem. Find all monotonic functions $F(x)$ for which there exist auxiliary functions $A(a)$ and $B(a)$ for which

$$F(a \cdot x) = A(a) \cdot F(x) + B(a)$$

for all x and a.

Now, we are ready to formulate the main result of this section.

Observation. One can easily check that if a function $F(x)$ satisfies the desired property, then, for every two real numbers $c_1 > 0$ and c_0, the function $\overline{F}(x) \stackrel{\text{def}}{=} c_1 \cdot F(x) + c_0$ also satisfies this property. We thus say that the function $\overline{F}(x) = c_1 \cdot F(x) + c_0$ is *equivalent* to the original function $F(x)$.

Proposition 3.2.1. *Every monotonic solution of the above functional equation is equivalent to* $\log(x)$ *or to* x^{α}.

Conclusion. So, symmetries and similarities do explain the selection of the function $F(x)$ for I-complexity.

Proof of Proposition 3.2.1.

$1°$. Let us first prove that the desired function $F(x)$ is differentiable.

Indeed, it is known that every monotonic function is almost everywhere differentiable. Let $x_0 > 0$ be a point where the function $F(x)$ is differentiable. Then, for every x, by taking $a = x/x_0$, we conclude that $F(x)$ is differentiable at this point x as well.

$2°$. Let us now prove that the auxiliary functions $A(a)$ and $B(a)$ are also differentiable.

Indeed, let us pick any two real numbers $x_1 \neq x_2$. Then, for every a, we have $F(a \cdot x_1) = A(a) \cdot F(x_1) + B(a)$ and $F(a \cdot x_2) = A(a) \cdot F(x_2) + B(a)$. Thus, we get a system of two linear equations with two unknowns $A(a)$ and $B(a)$.

$$F(a \cdot x_1) = A(a) \cdot F(x_1) + B(a).$$

$$F(a \cdot x_2) = A(a) \cdot F(x_2) + B(a).$$

Based on the known formula (Cramer's rule) for solving such systems, we conclude that both $A(a)$ and $B(a)$ are linear combinations of differentiable functions $F(a \cdot x_1)$ and $F(a \cdot x_2)$. Hence, both functions $A(a)$ and $B(a)$ are differentiable.

$3°$. Now, we are ready to complete the proof.

Indeed, based on Parts 1 and 2 of this proof, we conclude that

$$F(a \cdot x) = A(a) \cdot F(x) + B(a)$$

for differentiable functions $F(x)$, $A(a)$, and $B(a)$. Differentiating both sides by a, we get

$$x \cdot F'(a \cdot x) = A'(a) \cdot F(x) + B'(a).$$

In particular, for $a = 1$, we get $x \cdot \dfrac{dF}{dx} = A \cdot F + B$, where $A \stackrel{\text{def}}{=} A'(1)$ and $B \stackrel{\text{def}}{=} B'(1)$.

So, $\dfrac{dF}{A \cdot F + b} = \dfrac{dx}{x}$; now, we can integrate both sides.

Let us consider two possible cases: $A = 0$ and $A \neq 0$.

$3.1°$. When $A = 0$, we get $\dfrac{F(x)}{b} = \ln(x) + C$, so

$$F(x) = b \cdot \ln(x) + b \cdot C.$$

$3.2°$. When $A \neq 0$, for $\widetilde{F} \stackrel{\text{def}}{=} F + \dfrac{b}{A}$, we get $\dfrac{d\widetilde{F}}{A \cdot \widetilde{F}} = \dfrac{dx}{x}$, so $\dfrac{1}{A} \cdot \ln(\widetilde{F}(x)) = \ln(x) + C$, and $\ln(\widetilde{F}(x)) = A \cdot \ln(x) + A \cdot C$. Thus, $\widetilde{F}(x) = C_1 \cdot x^A$, where $C_1 \stackrel{\text{def}}{=} \exp(A \cdot C)$. Hence, $F(x) = \widetilde{F}(x) - \dfrac{b}{A} = C_1 \cdot x^A - \dfrac{b}{A}$.

The proposition is proven.

3.3 How Can We Predict?

Practical Need to Find Dependencies. In situations where prediction is, in principle, possible, the next questions is: *how* can we predict? In cases where we know the corresponding equations, we can use these equations for prediction. In practice, it often occurs that we know (or conjecture) that a quantity y depends on quantities $x_1, \ldots, x_n$, but we do not know the exact form of this dependence. In such situations, we must experimentally determine this dependence $y = f(x_1, \ldots, x_n)$.

For that, in several (S) situations $s = 1, \ldots, N$, we measure the values of both the dependent variable y and of the independent variables x_i. Then, we use the results $\left(x_1^{(s)}, \ldots, x_n^{(s)}, y^{(s)} \right)$ of these measurements to find a function $f(x_1, \ldots, x_n)$ which is consistent with all these measurement results, i.e., for which

$$y^{(s)} \approx f\left(x_1^{(s)}, \ldots, x_n^{(s)} \right)$$

for all s from 1 to S. (The equality is usually approximate since the measurements are approximate and the value y is often only approximately determined by the values of the variables $x_1, \ldots, x_n$.)

First Approximation: Linear Dependence. In many practical situations, the dependence $f(x_1, \ldots, x_n)$ is smooth: informally, this means that small changes in x_i lead to equally small changes in y. In the first approximation, a smooth function can be approximated by its tangent, i.e., by a linear expression

$$f(x_1, \ldots, x_n) = c + \sum_{i=1}^{n} c_i \cdot x_i$$

for appropriate coefficients c and c_i.

The task of estimating the values of these coefficients is based on the measurement results $\left(x_1^{(s)}, \ldots, x_n^{(s)}, y^{(s)} \right)$, i.e., based on the system of equations

$$y^{(s)} \approx c + \sum_{i=1}^{n} c_i \cdot x_i^{(s)},$$

is known as *linear regression*; see, e.g., [143].

Need to Go beyond Linear Dependencies. To get a more accurate description of the desired dependence, we need to go beyond the first (linear) approximation.

A Natural Mathematical Approach. A natural mathematical idea – traditionally used in statistical analysis – is that since the first (linear) approximation does not work well, we need to use the second (quadratic) approximation. In other words, we need to describe the desired dependence as

$$f(x_1, \ldots, x_n) = c + \sum_{i=1}^{n} c_i \cdot x_i + \sum_{i=1}^{n} \sum_{j=1}^{n} c_{ij} \cdot x_i \cdot x_j$$

for appropriate coefficients c, c_i, and c_{ij}. Statistical regression methods enable us to find the coefficients from the corresponding system of linear equations:

$$y^{(s)} \approx c + \sum_{i=1}^{n} c_i \cdot x_i^{(s)} + \sum_{i=1}^{n} \sum_{j=1}^{n} c_{ij} \cdot x_i^{(s)} \cdot x_j^{(s)}.$$

A Neural Network Approach Is Often More Efficient. In practice, often, it is more computationally efficient to use neural networks; see, e.g., [13].

In the traditional (3-layer) neural networks, the input values $x_1, \ldots, x_n$:

- first go through the non-linear layer of "hidden" neurons, resulting in the values

$$y_i = s_0 \left(\sum_{j=1}^{n} w_{ij} \cdot x_j - w_{i0} \right), \quad 1 \le i \le m,$$

- after which a linear neuron combines the results y_i into the output

$$y = \sum_{i=1}^{m} W_i \cdot y_i - W_0.$$

Here, W_i and w_{ij} are *weights* selected based on the data, and $s_0(x)$ is a non-linear *activation function*. Usually, the "sigmoid" activation function is used:

$$s_0(x) = \frac{1}{1 + \exp(-x)}.$$

The weights W_i and w_{ij} are selected so as to fit the data, i.e., that $y^{(s)} \approx f\left(x_1^{(s)}, \ldots, x_n^{(s)}\right)$ for all $s = 1, \ldots, S$.

A Natural Question. A natural question is: why are neural networks – when appropriately applied – a more computationally efficient approximation? In this section, we provide an explanation for this empirical phenomenon. This explanation first appeared in [109].

Why Symmetries. At first glance, the use of symmetries in neural networks may sound somewhat strange, because there are no *explicit* symmetries there, but *hidden* symmetries have been actively used in neural networks. For example, they are the only way to explain the empirically observed advantages of the sigmoid activation function; see, e.g., [78, 119].

Symmetry and Similarity: A Fundamental Property of the Physical World. One of the main objectives of science is prediction. What is the usual basis for prediction? We observed similar situations in the past, and we expect similar outcomes. In mathematical terms, similarity corresponds to *symmetry*, and similarity of outcomes – to *invariance*.

For example, we dropped the ball, it fell down. We conclude that if we drop it at a different location and/or at a different orientation, it will also fall down. Why – because we believe that the process is invariant with respect to shifts, rotations, etc.

This fundamental role of symmetries and similarities is well recognized in modern physics, to the extent that, starting with the quark theory, theories are usually formulated in terms of the corresponding symmetries – and not in terms of differential equations as it was in Netwon's time and later. Of course, once the symmetries are known, we can determine the equations, but they are no longer the original formulation.

It is therefore natural to apply symmetries to neural networks as well.

Basic Symmetries: Scaling and Shift. What are the basic symmetries? As we have mentioned in Chapter 1, typically, we deal with the numerical values of a physical quantity. Numerical values depend on the *measuring unit*. If we use a new unit which is λ times smaller, numerical values are multiplied by λ: $x \to \lambda \cdot x$. For example, x meters $= 100 \cdot x$ cm. The transformation $x \to \lambda \cdot x$ is usually called *scaling*.

Another possibility is to change the starting point. For example, instead of measuring time from year 0, we can start measuring it from some more distant year in the past. If we use a new starting point which is s units smaller, then the quantity which was originally represented by the number x is now represented by the new value $x + s$. The transformation $x \to x + s$ is usually called a *shift*.

So, we arrive at the following natural requirement: that the physical formulas should not depend on the choice of a measuring unit or of a starting point. Together, scaling and shifts form *linear transformations* $x \rightarrow a \cdot x + b$. Thus, in mathematical terms, this means that the physical formulas be invariant under linear transformations.

Basic Nonlinear Symmetries. Sometimes, a system also has *nonlinear* symmetries. To find such non-linear symmetries, we can take into account that if a system is invariant under f and g, then:

- it is invariant under their composition $f \circ g$, and
- it is invariant under the inverse transformation f^{-1}.

In mathematical terms, this means that symmetries form a *group*.

In practice, at any given moment of time, we can only store and describe finitely many parameters. Thus, it is reasonable to restrict ourselves to *finite-dimensional* groups.

One of the first researcher to explore this idea was Norbert Wiener, the father of cybernetics. He formulated a question: describe all finite-dimensional groups that contain all linear transformations. For transformations from real numbers to real numbers, the answer to this question are known: all elements of this group are fractionally-linear functions $x \rightarrow \dfrac{a \cdot x + b}{c \cdot x + d}$.

Symmetries and Similarities Explain the Choice of an Activation Function. Let us show that such non-linear symmetries explain the formula for the *activation function* $f(x) = \dfrac{1}{1 + \exp(-x)}$.

Indeed, a change in the input starting point has the form $x \rightarrow x + s$. It is reasonable to require that the new output $f(x + s)$ is equivalent to the $f(x)$ modulo an appropriate transformation. We have just mentioned that appropriate transformations are fractionally linear. Thus, we conclude that for every s, there exist values $a(s)$, $b(s)$, $c(s)$, and $d(s)$ for which

$$f(x+s) = \frac{a(s) \cdot f(x) + b(s)}{c(s) \cdot f(x) + d(s)}.$$

Differentiating both sides by s and equating s to 0, we get a differential equation for $f(x)$. Its known solution is the sigmoid activation function – which can thus be explained by symmetries and similarities.

Apolloni's Idea. One of the problems with the traditional neural networks is that in the process of learning – i.e., in the process of adjusting the values of the weights to fit the data – some of the neurons are duplicated, i.e., we get $w_{ij} = w_{i'j}$ for some $i \neq i'$ and thus, $y_i = y_{i'}$.

As a result, we do not fully use the learning capacity of a neural network, since when $y_i = y_{i'}$, we can get the same approximation with fewer hidden neurons.

To avoid the above redundancy problem, B. Apolloni and others suggested [2] that we *orthogonalize* the neurons during training, e.g., that we make sure that the corresponding linear combinations $\sum_{j=1}^{n} w_{ij} \cdot x_j$ remain orthogonal in the sense that

$$\langle w_i, w_{i'} \rangle \stackrel{\text{def}}{=} \sum_{j=1}^{n} w_{ij} \cdot w_{i'j} = 0$$

for all $i \neq i'$. where we denoted $w_i = (w_{i1}, \ldots, w_{in})$.

Challenge. Since Apolloni *et al.* heuristic idea works well, it is desirable to look for its precise mathematical justification. In this section, we provide such a justification in terms of symmetries and similarities.

Comment. This result first appeared in [106, 110].

Towards Formulating the Problem in Precise Terms. We must select a basis $e_0(x)$, $e_1(x)$, ..., $e_n(x)$, ... so that each function $f(x)$ is represented as $f(x) = \sum_i c_i \cdot e_i(x)$. For example:

- an expansion in Taylor series corresponds to choosing the basis $e_0(x) = 1$, $e_1(x) = x$, $e_2(x) = x^2$, ...
- an expansion in Fourier series corresponds to selecting the basis

$$e_i(x) = \sin(\omega_i \cdot x).$$

Once the basis is selected, to store the information about the function $f(x)$, we store the coefficients $c_0, c_1, \ldots$, corresponding to this basis.

From this viewpoint, one of the possible criteria for selecting the basis can be that the selected basis should require, on average, the smallest number of bits to store $f(x)$ with given accuracy. We can think of several similar criteria.

For all these criteria, we can take into account that storing a number c_i and storing the opposite number $-c_i$ take the same space. Thus, changing one of the basis function $e_i(x)$ to $e'_i(x) = -e_i(x)$ (which we lead to exactly this change $c_i \rightarrow -c_i$) does not change accuracy or storage space. So, we conclude that:

- if $e_0(x), \ldots, e_{i-1}(x), e_i(x), e_{i+1}(x), \ldots$ is an optimal basis,
- then the basis $e_0(x), \ldots, e_{i-1}(x), -e_i(x), e_{i+1}(x), \ldots$ is also optimal.

Uniqueness of the Optimal Solution. Due to the previous argument, we do not select a single basis, we select a family $\pm e_0(x), \pm e_1(x), \ldots$, in which each function is determined modulo its sign. Out of all such families, we should select the optimal one.

In general, an optimization problem may have several optimal solutions. In this case, we can use this non-uniqueness to optimize something else. For example:

- if two sorting algorithms are equally fast in the worst case $t^w(A) = t^w(A')$,
- we can select the one with the smallest average time $t^a(A) \rightarrow \min$.

In effect, by introducing the additional criterion, we now have a new criterion: A is better than A' if either $t^w(A) < t^w(A')$ or $(t^w(A) = t^w(A')$ and $t^a(A) < t^a(A'))$.

If this new criterion also has several optimal solutions, we can optimize something else, etc., until we end up with a unique optimal solution. So, non-uniqueness means that the original criterion was not final. Relative to a *final* criterion, there is *only one* optimal solution.

For our problem, this uniqueness means that:

- once we have one optimal basis

$$e_0(x), \quad e_1(x), \quad e_2(x), \quad \ldots,$$

- all other optimal bases have the form

$$\pm e_0(x), \quad \pm e_1(x), \quad \pm e_2(x), \quad \ldots$$

How to Describe Average Accuracy. Our objective is to describe average accuracy, or average number of bits, etc. We also want to know the mean square distance $\int (f(x) - f_\approx(x))^2 \, dx$ between the original function $f(x)$ and its approximation $f_\approx(x)$.

To describe these averages, we need to know the corresponding probability distribution on the set of all possible functions $f(x)$.

Dependencies $f(x)$ come from many different factors. Due to Central Limit Theorem, it is thus reasonable to assume that the distribution on $f(x)$ is Gaussian. If $m(x) \overset{\text{def}}{=} E[f(x)] \neq 0$, we can store differences $\Delta f(x) \overset{\text{def}}{=} f(x) - m(x)$, for which $E[\Delta f(x)] = 0$. Thus, without losing generality, we can assume that $E[f(x)] = 0$.

Such Gaussian distributions are uniquely determined by their covariances $C(x,y) \overset{\text{def}}{=} E[f(x) \cdot f(y)]$. A general Gaussian distribution can be described by independent components: $f(x) = \sum_i \eta_i \cdot f_i(x)$, where $E[\eta_i \cdot \eta_j] = 0$, $i \neq j$. The corresponding functions $f_i(x)$ are eigenfunctions of the covariance function $C(x,y) = E[f(x)f(y)]$:

$$\int C(x,y) \cdot f_j(y) \, dy = \lambda_j \cdot f_j(x).$$

The basis formed by these functions is known as the *Kahrunen-Loeve* (KL) basis. The functions from the KL basis together with the corresponding eigenvalues λ_i uniquely determine the corresponding probability distribution – and thus, the value of the optimality criterion.

Functions from this *KL basis* are orthogonal; they are usually selected to be orthonormal, i.e., satisfy the condition $\int f_j^2(x) \, dx = 1$.

In the general case, where all eigenvalues λ_j are different, each eigenfunction $f_j(x)$ is determined uniquely modulo $f_j(x) \to -f_j(x)$.

One can easily see that if we change one of functions $f_j(x)$ from the KL basis to $-f_j(x)$, we get a KL basis. Under this change, the values $E[f(x) \cdot f(y)]$ and $\int f^2(x) \, dx$ do not change – and thus, optimality criteria based on these values do not change. Thus, we arrive at the following formulation of the problem.

Formulation of the Problem in Precise Terms. We have an optimality criterion described in terms of a sequence of orthonormal functions $f_j(x)$ and a sequence of corresponding numbers λ_i. We know that functions $\pm f_j(x)$ determine the exact same criterion as the original functions $f_j(x)$.

We consider the generic case, in which all the eigenvalues λ_j are different.

Based on this criterion, we must select an optimal basis $e_0(x)$, $e_1(x)$, ..., $e_i(x)$, ...Each function from the desired basis can be represented as a linear combination of functions from the KL basis:

$$e_i(x) = \sum_j a_{ij} \cdot f_j(x).$$

Thus, selecting an optimal basis is equivalent to selecting the matrix of values a_{ij}, and the optimality criterion is equivalent to selecting a class of all matrices corresponding to optimal functions.

Of course, since the vectors $e_i(x)$ must form a basis, we cannot have $e_i(x) \equiv 0$, i.e., for every i, at least one value a_{ij} must e different from 0. We call such matrices *non-trivial*.

We have mentioned that if we change one of the functions $f_{j_0}(x)$ to $-f_{j_0}(x)$, the criterion does not change. Thus, the following functions also form an optimal basis:

$$e'_i(x) = \sum_{j \neq j_0} a_{ij} \cdot f_j(x) - a_{ij_0} \cdot f_{j_0}(x).$$

These functions correspond to the new matrix a'_{ij} for which $a'_{ij_0} = -a_{ij_0}$ and $a'_{ij} = a_{ij}$ for all $j \neq j_0$.

We also require that every optimal basis has the form $e'_i(x) = \pm e_i(x)$. Thus, we arrive at the following definition.

Definition 3.3.1. *Let $f_i(x)$ be a sequence of linearly independent functions.*

- *We say that a matrix a_{ij} is* non-trivial *if for every i, there exists a j for which $a_{ij} \neq 0$.*
- *By an* optimality criterion, *we mean a class A of non-trivial matrices a_{ij}.*
- *For each matrix $a_{ij} \in A$, the functions $e_i(x) = \sum_j a_{ij} \cdot f_j(x)$ are called* optimal *functions corresponding to this matrix.*
- *We say that the optimality criterion is* invariant *if for every matrix $a_{ij} \in A$ and for every j_0, the matrix a'_{ij}, for which $a'_{ij_0} = -a_{ij_0}$ and $a'_{ij} = a_{ij}$ for all $j \neq j_0$, also belongs to the class A.*
- *We say that the optimality criterion is* final *if for every two matrices $a_{ij}, a'_{ij} \in A$ and for every i, the corresponding optimal functions $e_i(x) = \sum_j a_{ij} \cdot f_j(x)$ and $e'_i(x) = \sum_j a'_{ij} \cdot f_j(x)$ differ only by sign, i.e., either $e'_i(x) = e_i(x)$ or $e'_i(x) = -e_i(x)$.*

Theorem 3.3.1. *If an optimality criterion is invariant and final, then each optimal function $e_i(x)$ has the form $e_i = a_{ij_0} \cdot f_{j_0}(x)$ for some j_0.*

Proof. Indeed, let a_{ij} be a matrix from the optimal criterion. Since the matrix is non-trivial, for every i, there exists a j_0 for which $a_{ij_0} \neq 0$. Since the optimality criterion is invariant, the class A also contains the matrix a'_{ij} for which $a'_{ij_0} = -a_{ij_0}$ and $a'_{ij} = a_{ij}$ for all $j \neq j_0$. For this new matrix, the corresponding optimal functions have the form

$$e'_i(x) = \sum_{j \neq j_0} a_{ij} \cdot f_j(x) - a_{ij_0} \cdot f_{j_0}(x).$$

Since the optimality criterion is final, this expression must be equal either to $e_i(x)$ or to $-e_i(x)$.

If $e'_i(x) = e_i(x)$, we would have

$$\sum_{j \neq j_0} a_{ij} \cdot f_j(x) - a_{ij_0} \cdot f_{j_0}(x) = \sum_{j \neq j_0} a_{ij} \cdot f_j(x) + a_{ij_0} \cdot f_{j_0}(x).$$

The difference between the two sides is equal to 0, hence $a_{ij_0} \cdot f_{j_0}(x) = 0$ and $a_{ij_0} = 0$, but we have selected j_0 for which $a_{ij_0} \neq 0$. Thus, $e'_i(x) = e_i(x)$ is impossible, so we must have $e'_i(x) = -e_i(x)$, i.e.,

$$\sum_{j \neq j_0} a_{ij} \cdot f_j(x) - a_{ij_0} \cdot f_{j_0}(x) = - \sum_{j \neq j_0} a_{ij} \cdot f_j(x) - a_{ij_0} \cdot f_{j_0}(x).$$

Since the functions $f_j(x)$ are linearly independent, this equality implies that the coefficients at all $f_j(x)$ in both sides must coincide. In particular, by comparing the coefficients at $f_j(x)$ for every $j \neq j_0$, we conclude that $a_{ij} = -a_{ij}$ hence $a_{ij} = 0$. So, $a_{ij} = 0$ for all $j \neq j_0$. The theorem is proven.

Discussion. We have proved that for the optimal basis $e_i(x)$ and for the KL basis $f_j(x)$, each $e_i(x)$ has the form

$$e_i(x) = a_{ij_0} \cdot f_{j_0}(x) \text{ for some } a_{ij_0}.$$

We know that the elements $f_j(x)$ of the KL basis are orthogonal. So, we conclude that the elements $e_i(x)$ of the optimal basis are orthogonal as well.

Apolloni's idea was to always make sure that we use an orthogonal basis. This idea has been empirically successful. Our new result provides a theoretical justification for Apolloni's idea.

Why Neural Networks Lead to a More Computationally Efficient Approximation? Now that we have explained how to get rid of the redundancy problem, let us return to the question that we formulated in the beginning of this section: why neural networks lead to a more computationally efficient approximation?

Apolloni's Idea: Reminder. To avoid the above redundancy problem, B. Apolloni and others suggested [2] that we *orthogonalize* the neurons during training, e.g., that we make sure that the corresponding linear combinations $\sum_{j=1}^{n} w_{ij} \cdot x_j$ remain orthogonal in the sense that

$$\langle w_i, w_{i'} \rangle \stackrel{\text{def}}{=} \sum_{j=1}^{n} w_{ij} \cdot w_{i'j} = 0$$

for all $i \neq i'$. where we denoted $w_i = (w_{i1}, \ldots, w_{in})$; see also [106, 110].

Neural Networks in the Second Approximation: Analysis. We consider the second approximation, in which each function is approximated by a quadratic expression – e.g., by the sum of the constant, linear, and quadratic terms of its Taylor expansion, so that cubic and higher orders can be safely ignored.

In the second approximation, we can approximate the non-linear activation function $s_0(x)$ by the sum of its constant, linear, and quadratic terms:

$$s_0(x) \approx s + s_1 \cdot x + s_2 \cdot x^2.$$

In this case, the above formula for the output of an intermediate neuron takes the following form:

$$y_i = s_0 + s_1 \cdot \left(\sum_{j=1}^{n} w_{ij} \cdot x_j - w_{i0} \right) + s_2 \cdot \left(\sum_{j=1}^{n} w_{ij} \cdot x_j - w_{i0} \right)^2.$$

The quadratic term in this expression can be described as

$$\left(\sum_{j=1}^{n} w_{ij} \cdot x_j - w_{i0} \right)^2 = \left(\sum_{j=1}^{n} w_{ij} \cdot x_j \right)^2 - 2w_{i0} \cdot \left(\sum_{j=1}^{n} w_{ij} \cdot x_j \right) + w_{i0}^2.$$

Here, the term

$$\left(\sum_{j=1}^{n} w_{ij} \cdot x_j \right)^2 = (\langle w_i, x \rangle)^2$$

is the only quadratic terms, the other terms are linear, where we denote $x = (x_1, \ldots, x_n)$. Thus, the output $y = \sum_{i=1}^{m} W_i \cdot y_i - W_0$ of the neural networks consists of a linear part plus a quadratic part of the type

$$Q_n = \sum_{i=1}^{n} W_i \cdot \langle w_i, x \rangle^2.$$

This part corresponds to the quadratic part $\sum_{i=1}^{n} \sum_{j=1}^{n} c_{ij} \cdot x_i \cdot x_j$ of the original Taylor-series representation:

$$Q_n = \sum_{i=1}^{n} W_i \cdot \langle w_i, x \rangle^2 = \sum_{i=1}^{n} \sum_{j=1}^{n} c_{ij} \cdot x_i \cdot x_j.$$

As we have mentioned, it is reasonable to select the vectors w_i to be orthogonal. By dividing each vector by its length (and appropriately multiplying W_i by this length), we can assume that the vectors are also orthonormal, i.e., that $\langle w_i, w_i \rangle = 1$ for all i. In the orthonormal basis formed by these vectors w_i,

- the corresponding matrix c_{ij} becomes a diagonal matrix,
- with values W_i on the diagonal.

Thus:

- the vectors w_i are eigenvectors of the matrix c_{ij}, while
- the values W_i are the eigenvalues of this matrix.

So, we arrive at the following conclusion.

Difference between Traditional Statistical Representation and Neural Network Representation Reformulated. In the second approximation, a generic non-linear part of a function can be represented by a general symmetric matrix c_{ij}. We consider the two competing representations of a function $f(x_1, \ldots, x_n)$:

- the traditional statistical representation in terms of the first few terms of Taylor series and
- a neural network representation.

In terms of the matrix c_{ij}, these two representations correspond to the following:

- in the traditional statistical representation, we store all the components c_{ij} of the original matrix;
- in the neural network representation, we store instead the eigenvectors and eigenvalues of this matrix.

Physical Analogy: A Comment. The above conclusion prompts a natural analogy with quantum physics; see, e.g., [38]. In quantum physics, from the mathematical viewpoint, an observable quantity can be described by a corresponding matrix c_{ij}. However, a more physically natural description is to describe possible values of this quantity – which are exactly eigenvalues of this matrix – and states in which this quantity has these exactly values, which are eigenvectors of the matrix. In this example, a representation via eigenvalues and eigenvectors is clearly intuitively preferable.

Towards Efficient Computations. Our objective is to find an expression that, given the inputs $x_1, \ldots, x_n$, generates the value $y = f(x_1, \ldots, x_n)$.

Which operations are the most efficient on modern computers? In numerical computations that form the bulk of modern high performance computer usage, the most time-consuming operation is the *dot product*, i.e., computing the $\langle a, b \rangle$ for given vectors a and b.

The prevalence of dot product makes sense from the mathematical viewpoint, since most numerical methods are based on linearization, and in the linear approximation, any function of n variables is approximated as $c + \sum_{i=1}^{n} c_i \cdot x_i$, i.e., as a constant plus a dot product between the vector of inputs and the vector of coefficients.

Not surprisingly, most computer speed-up innovations are aimed at computing the dot product faster – e.g., the multiply-accumulate operation which is an important part of digital signal processing or fused multiple-add operation which is now hardware supported on many modern computers; see, e.g., [54].

From this viewpoint, the way to speed up any computation is to reduce it to as few dot products as possible.

How to Efficiently Compute $f(x_1, \ldots, x_n)$ under Both Representations. Computing the value of the linear part requires computing exactly one dot product.

Computing the value of the traditional quadratic form requires $n+1$ dot products:

- first, we compute n dot products $c_i \stackrel{\text{def}}{=} \sum\limits_{j=1}^{n} c_{ij} \cdot x_j$ for $i = 1, \ldots, n$;
- then, to find the desired value of the quadratic form, we compute the dot product $\sum\limits_{i=1}^{n} c_i \cdot x_i$.

In the neural network representation, to compute the value with a certain accuracy, we can dismiss the terms corresponding to small eigenvalues W_i. As a result, instead of the original formula with n eigenvalues, we get a simplified formula with $n' < n$ eigenvalues:

$$Q_n \approx \sum_{i=1}^{n'} W_i \langle w_i, x \rangle^2.$$

From this representation, we can see that fewer than $n+1$ dot products are needed:

- first, we compute $n' < n$ dot products $z_i = \langle w_i, x \rangle$ corresponding to n' non-dismissed eigenvectors w_i;
- then, we perform a component-wise vector operation to compute the values $t_i = z_i \cdot z_i$; such vector operations are highly parallelizable and can be performed really fast on most modern computers; see, e.g., [54];
- finally, to find the desired result, we compute the dot product $\sum\limits_{i=1}^{n'} W_i \cdot t_i$.

Resulting Comparison. If n' is smaller than n, then indeed the neural network representation can lead to faster computations. This explains the empirical fact that in data processing, neural networks are often more efficient than more traditional statistical methods.

Comment. To "flesh out" this conclusion, we need to estimate to what extent the number n' of non-dismissed eigenvalues is smaller than the number n of all eigenvalues. This estimation is done in the remaining part of this section.

Number of Dismissed Eigenvalues: Semi-heuristic Statistical Analysis: Idea. The idea is to dismiss some eigenvalues because their contribution is small. Of course, the number of small eigenvalues depends on the matrix c_{ij}. We would like to know how many such eigenvalues are there *on average*. To formulate this question in precise terms, we need to describe a reasonable probability distribution on the set of all possible matrices.

Random Matrices: Motivation. In general, for each element c_{ij} of the matrix, we can have both positive and negative values. There are no reasons to expect positive values to be more probable than the negative ones or vice versa. In other words, the situation seems to be symmetric with respect to changing the sign. Thus, the expected value of the element c_{ij} should also be invariant with respect to this transformation. The only number that remains invariant when we change the sign is zero, so we conclude that the mean value of each component c_{ij} should be zero.

Similarly, there is no reason to assume that some of the elements have a different probability distribution; thus, we assume that they are identically distributed. Finally, there is no reason to assume that there is correlation between different elements. Thus, we assume that all the elements are independent. Thus, we arrive at the model in which all the elements are independent identically distributed random variables with mean 0 and a variance σ^2.

Eigenvalues of Random Matrices. For such random matrices, the distribution of their eigenvalues follows the *Marchenko-Pastur law*; see, e.g., [51, 94, 121]. To be more precise, this law describes the limit case of the following situation. We have an $m \times n$ random matrix X whose elements are independent identically distributed random variables with mean 0 and variance σ^2. Assume that m and n increase in such a way that the ratio m/n tends to a limit $\alpha > 0$. Then, for large n and m, the probability distribution of the eigenvalues of the matrix $Y = XX^T$ is asymptotically equivalent to

$$\rho(x) = \left(1 - \frac{1}{\alpha}\right) \cdot \delta(x) + \rho_c(x),$$

where $\delta(x)$ is Dirac's delta-function (i.e., the probability distribution which is located at the point 0 with probability 1), and $\rho_c(x)$ is different from 0 for $x \in [\alpha_-, \alpha_+]$, where $\alpha_\pm = \sigma^2 \cdot (1 \pm \sqrt{\alpha})^2$, and

$$\rho_c(x) = \frac{1}{2 \cdot \pi \cdot \sigma^2} \cdot \frac{\sqrt{(\alpha_+ - x) \cdot (x - \alpha_-)}}{\alpha \cdot x}.$$

In our case, matrices are square, so $m = n$, $\alpha = 1$ and thus, we have $\alpha_- = 0$, $\alpha_+ = 4\sigma^2$ and thus, the limit probability distribution takes the simplified form

$$\rho(x) = \frac{1}{2 \cdot \pi \cdot \sigma^2} \cdot \frac{\sqrt{(4\sigma^2 - x) \cdot x}}{x}.$$

Eigenvalues x of the matrix $Y = XX^T$ are squares of eigenvalues λ of the original matrix X: $x = \lambda^2$.

We are interested in small eigenvalues. For small eigenvalues, we have $x \ll \sigma$, so the above formula can be further simplified, into

$$\rho(x) \sim \frac{1}{2 \cdot \pi \cdot \sigma^2} \cdot \frac{\sqrt{4\sigma^2 \cdot x}}{x} = \frac{1}{2 \cdot \pi \cdot \sigma^2} \cdot \frac{2 \cdot \sigma \cdot \sqrt{x}}{x} = \frac{1}{\pi \cdot \sigma} \cdot \frac{1}{\sqrt{x}}.$$

The probability density ρ_λ for $\lambda = \sqrt{x}$ can thus be found as

$$\rho_\lambda = \frac{dp}{d\lambda} = \frac{dp}{dx} \cdot \frac{dx}{d\lambda}.$$

For $x = \lambda^2$, we get

$$\frac{dx}{d\lambda} = \frac{d(\lambda^2)}{d\lambda} = 2\lambda,$$

thus,

$$\rho_\lambda(\lambda) = \frac{1}{\pi \cdot \sigma} \cdot \frac{1}{\sqrt{x}} \cdot 2\lambda = \frac{1}{\pi \cdot \sigma} \cdot \frac{1}{\lambda} \cdot 2\lambda = \frac{2}{\pi \cdot \sigma}.$$

This expression for the probability density does not depend on λ at all. Thus, small eigenvalues have an approximately uniform distribution.

Heuristic Derivation of the Number of Eigenvalues That Can Be Safely Ignored. We would like to dismiss all the eigenvalues $\lambda_i = W_i$ whose absolute values are smaller than (or equal to) some small number $\delta > 0$. The overall contribution c of these eigenvalues is equal to

$$c = \sum_{i:|\lambda_i| \le \delta} W_i \cdot \langle w_i, x \rangle^2.$$

Since eigenvectors are orthonormal, the n values $\langle w_i, x \rangle^2$ add up to $\langle x, x \rangle^2$. In particular, for unit vectors x, these n values add up to 1. It is reasonable to assume that values corresponding to different eigenvalues are similarly distributed. Under this assumption, all these values have the same mean. The sum of n such means is equal to 1, so each mean is equal to $1/n$.

Each value W_i can be positive or negative. It is reasonable to assume that both negative and positive values are equally possible, so the mean value of each product $W_i \cdot \langle w_i, x \rangle^2$ is 0. Thus, the mean value of the sum is also 0.

Since $\langle w_i, x \rangle^2 \approx \frac{1}{n}$, the variance should be approximately equal to $W_i^2 \cdot \frac{1}{n^2}$. It is also reasonable to assume that the products $W_i \cdot \langle w_i, x \rangle^2$ corresponding to different eigenvalues are independent. Thus, the variance V_c of their sum c is equal to sum of their variances, i.e., to

$$V_c = \frac{1}{n^2} \cdot \sum_{i:|\lambda_i| \le \delta} W_i^2.$$

Since the mean is 0, and c is the sum of the large number of small independent components, it is reasonable to conclude, due to the Central Limit theorem, that it is approximately normally distributed; see, e.g., [143]. So, with probability 99.9%, all the values of this sum are located within the three sigma interval $[-3\sqrt{V_c}, 3\sqrt{V_c}]$.

Thus, the square root $\sqrt{V_c}$ is a good indication of the size of the dismissed terms. The size of the function itself can be similarly estimates as $\sqrt{V}$, where

$$V = \frac{1}{n^2} \cdot \sum_i W_i^2,$$

and the sum is taken over all eigenvalues. We want to make sure that the dismissed part does not exceed a given portion ε of the overall sum, i.e., that $\sqrt{V_c} \cdot \varepsilon \cdot \sqrt{V}$, or, equivalently, $V_c \leq \varepsilon^2 \cdot V^2$.

Within this constraint, we want to dismiss as many eigenvalues as possible; thus, we should not have $V_c \ll \varepsilon^2 \cdot V^2$, because then, we would be able to dismiss more terms. We should thus have $V_c \approx \varepsilon^2 \cdot V^2$. Because of the above expressions for V_c and for V, we therefore get an equivalent formula

$$\frac{1}{n^2} \cdot \sum_{i:|\lambda_i| \leq \delta} W_i^2 \approx \varepsilon^2 \cdot \frac{1}{n^2} \cdot \sum_i W_i^2.$$

Multiplying both sides by n^2, we can simplify this requirement into

$$\sum_{i:|\lambda_i| \leq \delta} W_i^2 \approx \varepsilon^2 \cdot \sum_i W_i^2.$$

Since the probability distribution of eigenvalues is described by the density function ρ_λ, and the total number of these eigenvalues is n, we have

$$\sum_i W_i^2 \approx n \cdot \int_{-\infty}^{\infty} \lambda^2 \cdot \rho_\lambda(\lambda) d\lambda$$

and similarly,

$$\sum_{i:|\lambda_i| \leq \delta} W_i^2 \approx n \cdot \int_{-\delta}^{\delta} \lambda^2 \cdot \rho_\lambda(\lambda) d\lambda.$$

Thus, the above requirement takes the form

$$n \cdot \int_{-\delta}^{\delta} \lambda^2 \cdot \rho_\lambda(\lambda) d\lambda \approx \varepsilon^2 \cdot n \cdot \int_{-\infty}^{\infty} \lambda^2 \cdot \rho_\lambda(\lambda) d\lambda.$$

Dividing both sides by n, we can simplify this into

$$\int_{-\delta}^{\delta} \lambda^2 \cdot \rho_\lambda(\lambda) d\lambda \approx \varepsilon^2 \cdot \int_{-\infty}^{\infty} \lambda^2 \cdot \rho_\lambda(\lambda) d\lambda.$$

For small λ, as we have derived, $\rho_\lambda \approx \text{const}$, so

$$\int_{-\delta}^{\delta} \lambda^2 \cdot \rho_\lambda(\lambda) d\lambda \approx \int_{-\delta}^{\delta} \lambda^2 \cdot \text{const} \, d\lambda = \text{const} \cdot \delta^3$$

(for a slightly different constant, of course).

Thus, the above requirement takes the form $\delta^3 \approx \text{const} \cdot \varepsilon^2$, i.e., $\delta \approx \varepsilon^{2/3}$.

Numerical Example. So, for example, for $\varepsilon \approx 10\% = 0.1$, we get $\delta \approx 0.1^{2/3} \approx 0.2$, so $\approx 20\%$ of all the eigenvalues can be safely ignored. As a result, we get a 20% decrease in computation time.

3.4 How Accurate Is Prediction? Quantifying the Uncertainty of the Corresponding Model

Need for Model Validation. Most physical models are approximate. It is therefore necessary to estimate the model accuracy by comparing the model's predictions with the experimental data. This estimation of the model accuracy is known as *model validation*.

Case Study: The Thermal Challenge Problem. As the main case study, we consider a benchmark thermal problem presented at the 2006 Sandia Validation Challenge Workshop [30, 37, 57, 58, 124, 134]. In this problem, we need to analyze temperature response $T(x,t)$ of a safety-critical device to a heat flux.

Specifically, a slab of metal (or other material) of a given thickness L is exposed to a given heat flux q. We know:

- the initial temperature $T_i = 25$ C, and
- an approximate model:

$$T(x,t) = T_i + \frac{q \cdot L}{k} \cdot \left[\frac{(k/\rho C_p) \cdot t}{L^2} + \frac{1}{3} - \frac{x}{L} + \frac{1}{2} \cdot \left(\frac{x}{L}\right)^2 - \right.$$

$$\left. \frac{2}{\pi^2} \cdot \sum_{n=1}^{6} \frac{1}{n^2} \cdot \exp\left(-n^2 \cdot \pi^2 \cdot \frac{(k/\rho C_p) \cdot t}{L^2}\right) \cdot \cos\left(n \cdot \pi \cdot \frac{x}{L}\right) \right].$$

We do not know a priori how accurate is the approximate model.

As for the thermal conductivity k and the volumetric heat capacity of the material ρC_p, we know their nominal values, and we have measured values of k and ρC_p for different specimens.

We also have the results of measuring temperature for several different specimens (which are, in general, different from the specimens for which we measure k and ρC_p). Specifically, for each specimen, we measure temperature at different moments of time.

Let Us Start with a Simplified Problem. To better describe our idea, let us start with a simplified version of this problem, in which we assume that for each specimen, the values of all parameters – including the thermal conductivity k and the volumetric heat capacity of the material ρC_p – are known exactly. We also assume that the actual temperatures are known exactly, i.e., that the temperature measurements are reasonably accurate – so that the measurement uncertainty can be safely ignored.

In this simplified situation, the predicted value $T(x,t)$ of the temperature is well defined for all x and t; the only reason why the measured values are different from the model's predictions is that the model itself is only approximate. So, to estimate the accuracy (or inaccuracy) in a model, we can simply compare these predictions $T(x,t)$ with the actual measurement results $\widetilde{T}(x,t)$.

The largest possible difference $\max\limits_{x,t} \left| \widetilde{T}(x,t) - T(x,t) \right|$ between the measured values and the theory's prediction can be used as a reasonable measure of the model's accuracy.

For example, if in all the measurements, the measured values differ from the theory's prediction by no more than 10 degrees, we conclude that the model's prediction are accurate with the accuracy ± 10 degrees.

How to Take Variability into Account: The Probabilistic Approach. In real life, the values of the parameters k and ρC_p are only approximately known. It is known that these values differ from one specimen to another. How can we take this variability into account when we estimate the accuracy of the given model?

A probabilistic approach to solving this problem is described in [37]. This approach is motivated by the fact that while we do not know the *individual* values of the parameters k and ρC_p corresponding to different specimens, we do have a *sample* of values k and ρC_p corresponding to different specimens. Thus, we can estimate the probability distribution of k and ρC_p among the given class of specimens.

In the resulting description, k and ρC_p are random variables with known distributions. Since the model's parameters k and ρC_p are random, for each x and t, the resulting temperature $T(x,t)$ also becomes a random variable. By running simulations, we can find, for each x and t, the probability distribution of this random value $T(x,t)$ – the probability distribution that would be observed if the model $T(x,t)$ was absolutely accurate.

Since the model is only approximately true, for every x and t, the actual (empirical) probability distribution of the measured temperatures $\widetilde{T}(x,t)$ is, in general, different from the simulated distribution of the model's predictions. The difference between these two probability distributions – the distribution predicted by the model and the distribution observed in measurements – can be thus viewed as a measure of how accurate is our model.

Limitation of the Probabilistic Approach: Description and Need to Overcome These Limitations. In the probabilistic approach, to describe an empirical distribution, we, in effect, combine ("pool") the temperatures measured for all the specimens into a single sample. As a result, we ignore an important part of the available information about the measurement results – namely, the information that some measurements correspond to the same specimen and some measurements correspond to different specimens. To get more convincing estimates of the model, it is therefore desirable to take this additional information into account.

In this section, we describe how this additional information can be used. We illustrate our approach on the example of the main case study. After that, we describe this approach in general terms, and provide another application example – Very Long Baseline Interferometry (VLBI).

Comment. In this section, we gauge the model's accuracy by coming up with a guaranteed upper bound for the difference between the model's prediction and actual values. This approach is similar to using overall error bound Δ as a description of the measurement inaccuracy – i.e., the difference between the measurement result $\widetilde{x}$ and the actual value x; see, e.g., [136]. In measurements, once we have the measurement result $\widetilde{x}$ and the bound Δ for which $|\widetilde{x} - x| \leq \Delta$, the only information that we have about the actual (unknown) values x is that x belongs to the interval $[\widetilde{x} - \Delta, \widetilde{x} + \Delta]$; see, e.g., [96]. Because of this similarity, we call our approach *interval approach*. This approach first appeared in [111].

What We Know about Each Specimen: An Example. Instead of pooling all the measured temperature values corresponding to different specimens into a single sample, we would like to consider each specimen individually. For each specimen, we have temperatures measured at different moments of time.

For example, according to [30], we have several specimens corresponding to Configuration 1, in which the thickness L is equal to 1.27 cm (half an inch), and the heat flux is equal to $q = 1000$ W/m^2. We have the measurement results for four specimens corresponding to this configuration. These measurement results correspond to $x = 0$. The results of measuring the temperature $T(x,t) = T(0,t)$ for specimen i are known as Experiment i. In particular, the measurement results corresponding to specimen 1 (i.e., to Experiment 1) are as shown in Table 3.4.1.

Table 3.4.1 Measurement results

time (in sec)	measured temperature
100	105.5
200	139.3
300	165.5
400	188.7
500	210.6
600	231.9
700	253.0
800	273.9
900	294.9
1000	315.8

Ideal Case: Exact Model, Exactly Known Parameters k and ρC_p. For each specimen, if the model was absolutely accurate (and if the measurement inaccuracy was negligible), the measured values $\widetilde{T}(x,t)$ would take the form $\widetilde{T}(x,t) = T(x,t,k,\rho C_p)$ for an appropriate values k and ρC_p; here, $T(x,t,k,\rho C_p)$ means that we explicitly take into account the dependence on the parameters k and ρC_p in the above formula.

In this ideal situation, if we know the exact values of k and ρC_p, to check the model's correctness, we can simply compare the measured values $\widetilde{T}(x,t)$ with the predicted values $T(x,t,k,\rho C_p)$. In this case, the largest possible difference between the measured and predicted values is 0: $\max_t \left| \widetilde{T}(x,t) - T(x,t,k,\rho C_p) \right| = 0$. Vice versa, if this largest difference is equal to 0, this means that all the differences are equal to 0, i.e., that the model is indeed absolute accurate.

Case Where the Model Is Exact, but the Parameters k and ρC_p Are Only Approximately Known. In reality, we do not know the exact values of k of ρC_p, so we can only conclude that this largest difference is equal to 0 for *some* values k and ρC_p. In other words, we conclude that the smallest possible value of this largest difference – smallest over all possible combinations of the parameters k and ρC_p – is equal to 0:

$$\min_{k,\rho C_p} \max_t \left| \widetilde{T}(x,t) - T(x,t,k,\rho C_p) \right| = 0.$$

Vice versa, if this smallest value is equal to 0, this means that for *some* k and ρC_p, the largest error $\max_t \left| \widetilde{T}(x,t) - T(x,t,k,\rho C_p) \right|$ is equal to 0 and so, the model is absolutely accurate.

General Case, Where We Take into Account That the Model Is Approximate. In practice, the model is approximate. This means that no matter which values k and ρC_p we use for this specimen, the measured values are different from the model's prediction: $\max_t \left| \widetilde{T}(x,t) - T(x,t,k,\rho C_p) \right| > 0.$

For example, if the model differs from the observations by some value $\varepsilon > 0$, then even for the actual values of k and ρC_p, we get $\left| \widetilde{T}(x,t) - T(x,t,k,\rho C_p) \right| = \varepsilon > 0$ and therefore, $\max_t \left| \widetilde{T}(x,t) - T(x,t,k,\rho C_p) \right| = \varepsilon > 0$. Moreover, even when the model differs from the actual values at a single moment t, we still have

$$\left| \widetilde{T}(x,t) - T(x,t,k,\rho C_p) \right| = \varepsilon > 0$$

for this moment of time t and therefore, $\max_t \left| \widetilde{T}(x,t) - T(x,t,k,\rho C_p) \right| = \varepsilon > 0.$

To gauge the accuracy of the model, it is therefore reasonable to use the difference corresponding to the best possible values k and ρC_p, i.e., the value

$$a \overset{\text{def}}{=} \min_{k,\rho C_p} \max_t \left| \widetilde{T}(x,t) - T(x,t,k,\rho C_p) \right|.$$

Comment. This difference a is observed when we use the exact values of the parameters k and ρC_p. If, for prediction, we use approximate values $\widetilde{k}$ and $\widetilde{\rho C_p}$, then, in addition to the inaccuracy ε of the model, we also have an additional inaccuracy caused by the inaccuracy in k and ρC_p. In this case, it is reasonable to expect that the worst-case difference between the observed and the predicted values is even larger than a:

$$\max_t \left| \widetilde{T}(x,t) - T(x,t,\widetilde{k},\widetilde{\rho C_p}) \right| > a.$$

Resulting Estimation of the Model's Accuracy: From the Analysis of a Single Specimen to the Analysis of all Measurement Results. For each specimen s, based on the observed values $\widetilde{T}_s(x,t)$ corresponding to this specimen, we can estimate the model's accuracy a_s in describing this specimen as

$$a_s = \min_{k,\rho C_p} \max_t \left| \widetilde{T_s}(x,t) - T_s(x,t,k,\rho C_p) \right|.$$

A model may have different accuracy for different specimens: e.g., a model may be more accurate for smaller values of the thermal flux q and less accurate for larger values of q. We are interested in guaranteed estimates of the model's accuracy, estimates which are applicable to all the specimens. Thus, as a reasonable estimate for the model's accuracy, we can take the largest value of a_s corresponding to different specimens:

$$a = \max_s a_s = \max_s \min_{k,\rho C_p} \max_t \left| \widetilde{T_s}(x,t) - T_s(x,t,k,\rho C_p) \right|.$$

Comment. The resulting formula for model's accuracy looks somewhat complicated, this is why we provided a detailed explanation of why we believe that this formula is adequate for model validation.

Estimating a_s as a Constrained Optimization Problem. The above formula for a_s means that we need to find the values k and ρC_p for which the difference $\left| \widetilde{T_s}(x,t) - T_s(x,t,k,\rho C_p) \right|$ is the smallest possible. In other words, for each specimen s, we want to minimize a_s under the constraints that

$$\widetilde{T_s}(x,t) - a_s \leq T_s(x,t,k,\rho C_p) \leq \widetilde{T_s}(x,t) + a_s$$

for all the measurement results $\widetilde{T_s}(x,t)$ obtained for this specimen.

Linearization as a First Approximation to This Constrained Optimization Problem. The dependence of the model prediction $T_s(x,t,k,\rho C_p)$ on the model prediction is non-linear. As a result, we get a difficult-to-solve non-linear optimization problem.

In practice, this problem can be simplified, because we know the nominal values $\widetilde{k}$ and $\widetilde{\rho C_p}$ of the parameters k and ρC_p, and we also know – from measurements – that the actual values of these parameters do not deviate too much from the nominal values: the differences $\Delta k = \widetilde{k} - k$ and $\Delta(\rho C_p) = \widetilde{\rho C_p} - \rho C_p$ are small. Thus, we can use the nominal values as the starting (0-th) approximations to k and ρC_p: $k^{(0)} = \widetilde{k}$ and $\rho C_p^{(0)} = \widetilde{\rho C_p}$.

In the first approximation, we can only keep terms which are linear in Δk and $\Delta(\rho C_p)$ in the expansion of the dependence

$$T_s(x,t,k,\rho C_p) = T_s\left(x,t,k^{(0)} - \Delta k, \rho C_p^{(0)} - \Delta(\rho C_p)\right):$$

$$T_s(x,t,k,\rho C_p) = T_s\left(x,t,k^{(0)},\rho C_p^{(0)}\right) - c_k^{(0)} \cdot \Delta k - c_{\rho C_p}^{(0)} \cdot \Delta(\rho C_p),$$

where

$$c_k^{(0)} \overset{\text{def}}{=} \frac{\partial T}{\partial k}, \quad c_{\rho C_p}^{(0)} \overset{\text{def}}{=} \frac{\partial T}{\partial(\rho C_p)},$$

and the derivatives are taken for $k = k^{(0)}$ and $\rho C_p = \rho C_p^{(0)}$. In this linear approximation, the above optimization problem takes the following form: minimize a_s under the constraints that

$$\widetilde{T}_s(x,t) - a_s \leq T_s\left(x,t,k^{(0)},\rho C_p^{(0)}\right) - c_k^{(0)} \cdot \Delta k - c_{\rho C_p}^{(0)} \cdot \Delta(\rho C_p) \leq \widetilde{T}_s(x,t) + a_s.$$

In this linearized problem, both the objective function and the constraints are linear in terms of unknowns, so we can use known (and efficient) algorithms of linear programming to solve this problem; see, e.g., [151].

Once we solve this problem, we get the values $\Delta k^{(1)}$ and $\Delta(\rho C_p)^{(1)}$ which are optimal in the first approximation. Based on these values, we can get a first approximation $k^{(1)}$ and $\rho C_p^{(1)}$ to the actual optimal values of k and ρC_p as $k^{(1)} = k^{(0)} - \Delta k^{(1)}$ and $\rho C_p^{(1)} = \rho C_p^{(0)} - \Delta(\rho C_p)^{(1)}$.

From a Linearized Solution to a General Solution. To get a more accurate solution, we can use the "approximately optimal" values $\Delta k^{(1)}$ and $\Delta \rho C_p^{(1)}$ as a new first approximation, and use linearization around these values. As a result, we arrive at the following iterative algorithm:

- we start with the values $k^{(0)} = \widetilde{k}$ and $\rho C_p^{(0)} = \widetilde{\rho C_p}$;
- on each iteration q, once we have the values $k^{(q-1)}$ and $\rho C_p^{(q-1)}$, we use linear programming to solve the following optimization problem: minimize a_s under the constraints that

$$\widetilde{T}_s(x,t) - a_s \leq T_s\left(x,t,k^{(q-1)},\rho C_p^{(q-1)}\right) - c_k^{(q-1)} \cdot \Delta k - c_{\rho C_p}^{(q-1)} \cdot \Delta(\rho C_p) \leq \widetilde{T}_s(x,t) + a_s,$$

where

$$c_k^{(q-1)} \overset{\text{def}}{=} \frac{\partial T}{\partial k}, \quad c_{\rho C_p}^{(q-1)} \overset{\text{def}}{=} \frac{\partial T}{\partial(\rho C_p)},$$

and the derivatives are taken for $k = k^{(q-1)}$ and $\rho C_p = \rho C_p^{(q-1)}$;
- once we solve this linear programming problem and get the optimal values $\Delta k^{(q)}$ and $\Delta(\rho C_p)^{(q)}$, we compute the next approximations to parameters as $k^{(q)} = k^{(q-1)} - \Delta k^{(q)}$ and $\rho C_p^{(q)} = \rho C_p^{(q-1)} - \Delta(\rho C_p)^{(q)}$.

Iterations continue until the process converges – or until we exhaust the computation time that was allocated for these computations. We then take the latest values of k and ρC_p and estimate the model's accuracy as $\max_g \widetilde{a}_g$, where

$$\widetilde{a}_g = \max_{x,t} \left| \widetilde{T}_s(x,t) - T_s(x,t,k,\rho C_p) \right|.$$

Numerical Results. For the above specimen 1, the iterative process converges after the 1st iteration (i.e., the 2nd iteration leads to very small changes). The resulting values of k and ρC_p lead to the predictions listed in Table 3.4.2.

Table 3.4.2 Prediction accuracy: interval approach

time (in sec)	measured temperature	prediction: interval approach
100	105.5	105.5
200	139.3	138.8
300	165.5	165.2
400	188.7	188.7
500	210.6	211.1
600	231.9	233.1
700	253.0	254.9
800	273.9	276.6
900	294.9	298.3
1000	315.8	319.9

The largest difference between the measured and predicted values is about 5 degrees. For other specimens, we got a similar difference of ≤ 5 degrees, so we conclude that the original model is accurate with accuracy ± 5 degrees.

How to Simplify Computations. To simplify computations, we used an equivalent reformulation of the original thermal model. Our main formula has the form

$$T(x,t) = T_i + \frac{q \cdot L}{k} \cdot \left[\frac{(k/\rho C_p) \cdot t}{L^2} + \frac{1}{3} - \frac{x}{L} + \frac{1}{2} \cdot \left(\frac{x}{L} \right)^2 - \right.$$

$$\left. \frac{2}{\pi^2} \cdot \sum_{n=1}^{6} \frac{1}{n^2} \cdot \exp\left(-n^2 \cdot \pi^2 \cdot \frac{(k/\rho C_p) \cdot t}{L^2} \right) \cdot \cos\left(n \cdot \pi \cdot \frac{x}{L} \right) \right].$$

In this formula, the parameter ρC_p always appears in a ratio $\dfrac{k/\rho C_p}{L^2}$. It is therefore reasonable, instead of the original variables $y_1 = k$ and $y_2 = \rho C_p$, to use new auxiliary variables $Y_1 = \dfrac{q \cdot L}{k}$ and $Y_2 = \dfrac{k/\rho C_p}{L^2}$. As a result, we get the following simplified formula:

$$T(x,t) = T_i + Y_1 \cdot \left[Y_2 \cdot t + \frac{1}{3} - x_0 + \frac{1}{2} \cdot x_0^2 - \right.$$

$$\left. \frac{2}{\pi^2} \cdot \sum_{n=1}^{6} \frac{1}{n^2} \cdot \exp(-n^2 \cdot \pi^2 \cdot Y_2 \cdot t) \cdot \cos(n \cdot \pi \cdot x_0) \right],$$

where $x_0 \overset{\text{def}}{=} \dfrac{x}{L}$. In this case,

$$\frac{\partial T}{\partial Y_1} = Y_2 \cdot t + \frac{1}{3} - x_0 + \frac{1}{2} \cdot x_0^2 - \frac{2}{\pi^2} \cdot \sum_{n=1}^{6} \frac{1}{n^2} \cdot \exp(-n^2 \cdot \pi^2 \cdot Y_2 \cdot t) \cdot \cos(n \cdot \pi \cdot x_0);$$

$$\frac{\partial T}{\partial Y_2} = t \cdot \left[Y_1 - 2 \cdot \sum_{n=1}^{6} \cdot \exp(-n^2 \cdot \pi^2 \cdot Y_2 \cdot t) \cdot \cos(n \cdot \pi \cdot x_0) \right].$$

Comments: how to get better accuracy estimates. The above model assumes that for each specimen, the values k and ρC_p remain the same. Measurement results show, however, that these values slightly change with temperature. This can be seen, e.g., if we plot the average value k_{av} of k measured for a given temperature as a function of temperature T; see Table 3.4.3.

Table 3.4.3 Dependence of k_{av} on T

T	20	250	500	750	1000
k_{av}	0.49	0.59	0.63	0.69	0.75

In the probabilistic approach, this dependence is taken into account by allowing correlation between the model and k; see, e.g., [37]. Linear correlation means, in effect, that instead of considering k as an independent random variables, we consider a dependence $k = k_0 + k_1 \cdot T$, where k_0 is independent on T and k_1 is a parameter to be determined. In the interval approach, for each specimen, we can similarly "plug in" the expressions $k = k_0 + k_1 \cdot T$ and $\rho C_p = \rho C_{p,0} + \rho C_{p,1} \cdot T$ into the above model and use the parameters k_0, k_1, $\rho C_{p,0}$, and $\rho C_{p,1}$ as the new unknowns in the similar constrained optimization approach.

Another possible improvement is related to the fact that we get slightly different values a_s depending on the thermal flow q: the higher q, the larger a_s. The objective is to predict how the system will react to thermal flows which may be even higher than in any of the experiments. So instead of taking the value $a(q_0)$ that corresponds to the current thermal flows q_0, we can estimate the dependence of $a(q)$ on q and extrapolate this dependence to the desired high thermal flow.

In our case, which model for the dependence $a(q)$ shall we choose? From the physical viewpoint, the problem is invariant w.r.t. changing measuring units $q \to \lambda \cdot q$ (i.e., in mathematical terms, scale-invariant). So it is reasonable to select a space-invariant dependence, i.e., a dependence for which, for each re-scaling $q \to \lambda \cdot q$, the dependence has the same form if we appropriate change the units for measuring a, i.e., that for every $\lambda > 0$, there exists a $C(\lambda)$ for which $a(\lambda \cdot q) = C(\lambda) \cdot a(q)$. It is known (see, e.g., [1]) that the only monotonic solutions to this functional equations have the form $a(q) = a_0 \cdot q^\alpha$ for some a_0 and α.

So, for each experimentally tested q, based on all samples with given q, we find $a(q)$, and then find a_0 and α for which $a(q) \approx a_0 \cdot q^\alpha$, i.e., equivalently, $\ln(a(q)) \approx \ln(a_0) + \alpha \cdot \ln(q)$. This is a system of linear equations with unknowns $\ln(a_0)$ and α, so we can use the Least Squares method to solve it. Once we find the solution, we can predict the model's accuracy as $a(q) \approx a_0 \cdot q^\alpha$.

Problem: General Description. In general, we have a model $z = f(x_1, \ldots, x_n, y_1, \ldots, y_m)$ that predicts the value z of the desired quantity as a function of known quantities $x_1, \ldots, x_n$ and unknown quantities $y_1, \ldots, y_m$; see, e.g., [126]. To be more precise, we usually know some crude approximate values $\widetilde{y}_i$, but the accuracy of these approximate values is orders of magnitude lower than the accuracy with which we know the measured values x_i and z.

Measurements are divided into groups with each of which we know that the values y_j are the same; the values y_j may differ from group to group.

Comment. In the thermal problem example, $n = 2$, $x_1 = x$, $x_2 = t$, $m = 2$, $y_1 = k$, and $y_2 = \rho C_p$. Groups correspond to specimens.

How to Estimate the Model's Accuracy: General Definition. In the general case, as an estimate for the model's accuracy, we propose to use the value

$$a = \max_{g} \min_{y_1, \ldots, y_m} \max_{x_1, \ldots, x_m} \left| \widetilde{f}_g(x_1, \ldots, x_n) - f_g(x_1, \ldots, x_n, y_1, \ldots, y_m) \right|,$$

where g indicates different groups, and $\widetilde{f}_g$ are measurement results corresponding to the g-th group.

In other words, as a desired value a, we take $\max_{g} a_g$, where each a_g is the solution to the following optimization problem: minimize a_g under the constraints that

$$\widetilde{f}_g(x_1, \ldots, x_n) - a_g \leq f_g(x_1, \ldots, x_n, y_1, \ldots, y_m) \leq \widetilde{f}_g(x_1, \ldots, x_n) + a_g.$$

How to Estimate the Model's Accuracy: General Algorithm. By applying a similar linearization approach, we get the following algorithm:

- we start with the values $z_i^{(0)} = \widetilde{z}_i$;
- on each iteration q, once we have the values $z_i^{(q-1)}$, we use linear programming to solve the following optimization problem: minimize a_g under the constraints that

$$\widetilde{f}_g(x_1, \ldots, x_n) - a_s \leq f_g \left(x_1, \ldots, x_n, y_1^{(q-1)}, y_m^{(q-1)} \right) - \sum_{j=1}^{m} c_j^{(q-1)} \cdot \Delta y_j \leq$$

$$\widetilde{f}_g(x_1, \ldots, x_n) - a_s,$$

where $c_j^{(q-1)} \overset{\text{def}}{=} \dfrac{\partial f}{\partial j}$, and the derivatives are taken for $y_j = y_j^{(q-1)}$;

- once we solve this linear programming problem and get the optimal values $\Delta y_j^{(q)}$, we compute the next approximations to parameters as

$$y_j^{(q)} = y_j^{(q-1)} - \Delta y_j^{(q)}.$$

Iterations continue until the process converges – or until we exhaust the computation time that was allocated for these computations. We then take the latest values of y_j and estimate the model's accuracy as $\max\limits_{g} \widetilde{a}_g$, where

$$\widetilde{a}_g = \max_{x_1,\ldots,x_n} \left| \widetilde{f}_g(x_1,\ldots,x_n) - f_g(x_1,\ldots,x_n,y_1,\ldots,y_m) \right|.$$

Very Long Baseline Interferometry (VLBI): Another Example of the General Approach. To get a better idea of the general problem, let us give another example of the general approach. For each distant astronomical radio-source, we want to find the exact direction from which the corresponding radio waves are coming. In precise terms, we need to find a unit vector $\mathbf{e}_k$ in the direction to the source.

One of the most accurate methods of finding the unit vector $\mathbf{e}_k$ in the direction to a distant astronomical radio-source is Very Long Baseline Interferometry (VLBI); see, e.g., [31, 32, 148, 153]. In VLBI, we measure the time delay $\tau_{i,j,k}$ between the signal observed by antennas i and j. The corresponding model comes the simple geometric arguments, according to which

$$\tau_{i,j,k} = c^{-1} \cdot (\mathbf{b}_i - \mathbf{b}_j) \cdot \mathbf{e}_k + \Delta t_i - \Delta t_j,$$

where:

- $\mathbf{b}_i$ is the location of the i-th antenna, and
- Δt_i is its clock bias on the i-th antenna, i.e., the difference between the reading of this clock and the actual (unknown) time on this antenna.

In this model, the locations $\mathbf{b}_i$ and the clock biases are unknown (to be more precise, we know approximate values of the locations and biases, but these approximate values are orders of magnitude less accurate that the time delays).

We assume that the directions $\mathbf{e}_k$ do not change during the measurements; this assumption make sense since the sources are distant ones, and even if they move with a speed v close to the speed of light, their angular speed v/R, where R is the distance, can be safely ignored. We also assume that the biases and the antenna locations do not change during one short group of measurements. In this case, z is the time delay, and $y_1,\ldots,y_m$ are directions $\mathbf{e}_k$, locations $\mathbf{b}_i$, and clock biases Δt_i. When we performed sufficiently many measurements in each group g, we have more measured values than the unknowns y_j and thus, we can meaningfully estimate the model's accuracy; for details, see [31, 32].

An even more accurate description emerges when we take into account that the Earth-bound antennas rotate with the Earth; to take rotation into account, we must take into account time between different consequent measurements within the same group, and this time can be measured very accurately – thus serving as x_i.

Closing Remarks. A model of real-life phenomena needs to be validated: we must compare the model's predictions with the experimental data and, based on this comparison, conclude how accurate is the model. This comparison becomes difficult if the model contains, as parameters, values of some auxiliary physical quantities

– quantities which are usually not measured in the corresponding experiments. In such situations, we can use the results of previous measurements of these quantities in similar situations, results based on which we can determine the probabilities of different values of these auxiliary quantities. In the traditional probabilistic approach to model validation, we plug in the resulting random auxiliary variables into the model, and compare the distribution of the results with the observed distribution of the experimental data. In this approach, however, we do use the important information that some measurement results correspond to the same specimen – and thus, correspond to the same values of the auxiliary quantities. To take this information into account, we propose a new approach, in which, for each specimen, we, in effect, first estimate the values of the auxiliary quantities based on the measurement results, then plug these estimated values back into the model – and use the resulting formula to gauge how accuracy the original model is on this specimen. We illustrate this approach on the example of a benchmark thermal problem.

3.5   Towards the Optimal Way of Processing the Corresponding Uncertainty

Functional Dependencies Are Ubiquitous. Several different quantities are used to describe the state of the world – or a state of a system in which are interested. For example, to describe the weather, we can describe the temperature, the wind speed, the humidity, etc. Even in simple cases, to describe the state of a simple mechanical body at a given moment of time, we can describe its coordinates, its velocity, its kinetic and potential energy, etc.

Some of these quantities can be directly measured – and sometimes, direct measurement is the only way that we can determine their values. However, once we have measured the values of a few basic quantities $x_1, \ldots, x_n$, we can usually compute the values of all other quantities y by using the known dependence $y = f(x_1, \ldots, x_n)$ between these quantities. Such functional dependencies are ubiquitous, they are extremely important in our analysis of real-world data.

Need for Polynomial Approximations. With the large amount of data that are constantly generated by different measuring devices, most of the data processing is performed by computers. So, we need to represent each known functional dependence $y = f(x_1, \ldots, x_n)$ in a computer.

In a computer, among operations with real numbers, the only ones which are directly hardware supported (and are therefore extremely fast) are addition, subtraction, and multiplication. All other operations with real numbers, including division, are implemented as a sequence of additions, subtractions, and multiplications.

Also hardware supported are logical operations; these operations make it possible to use several different computational expressions in different parts of the function domain. For example, computers use different approximation for trigonometric functions like $\sin(x)$ for $x \in [-\pi, \pi]$ and for inputs from the other cycles.

Therefore, if we want to compute a function $f(x_1,\ldots,x_n)$, we must represent it, on each part of the domain, as a sequence of additions, subtractions, and multiplications. A function which is obtained from variables $x_1,\ldots,x_n$ and constants by using addition, subtraction, and multiplication is nothing else but a polynomial. Indeed, one can easily check that every polynomial can be computed by a sequence of additions, subtractions, and multiplications. Vice versa, by induction, one can easily prove that every sequence of additions, subtractions, and multiplications leads to a polynomial; indeed:

- induction base is straightforward: each variables x_i is a polynomial, and each constant is a polynomial;
- induction step is also straightforward:

  - the sum of two polynomials is a polynomial;
  - the difference between two polynomials is a polynomial; and
  - the product of two polynomials is a polynomial.

Thus, we approximate functions by polynomials or by piece-wise polynomial functions (splines).

Possibility of a Polynomial Approximation. The possibility to approximate functions by polynomials was first proven by Weierstrass (long before computers were invented). Specifically, Weierstrass showed that for every continuous function $f(x_1,\ldots,x_n)$, for every box (multi-interval)

$$[a_1,b_1] \times \ldots \times [a_n,b_n],$$

and for every real number $\varepsilon > 0$, there exists a polynomial $P(x_1,\ldots,x_n)$ which is, on this box, ε-close to the original function $f(x_1,\ldots,x_n)$, i.e., for which

$$|P(x_1,\ldots,x_n) - f(x_1,\ldots,x_n)| \leq \varepsilon$$

for all $x_1 \in [a_1,b_1]$, $\ldots$, $x_n \in [a_n,b_n]$.

Polynomial (and piece-wise polynomial) approximations to a functional dependence have been used in science for many centuries, they are one of the main tools in physics and other disciplines. Such approximations are often based on the fact that most fundamental physical dependencies are analytical, i.e., can be expanded in convergent Taylor (polynomial) series. Thus, to get a description with a given accuracy on a given part of the domain, it is sufficient to keep only a few first terms in the Taylor expansion – i.e., in effect, to approximate the original function by a piece-wise polynomial expression; see, e.g. [38].

Polynomials Are Often Helpful. In many areas of numerical analysis, in particular, in computations with automatic results verification, it turns out to be helpful to approximate a dependence by a polynomial. For example, in computations with automatic results verification, Taylor methods – in which the dependence is approximated by a polynomial – turned out to be very successful; see, e.g., [10, 11, 12, 59, 91, 114].

The Efficiency of Polynomials Can Be Theoretically Explained. The efficiency of polynomials is not only an empirical fact, this efficiency can also be theoretically justified. Namely, in [115], it was shown that under reasonable assumptions on the optimality criterion – like invariance with respect to selection a starting point and a measuring unit for describing a quantity – every function from the optimal class of approximating functions is a polynomial.

How to Represent Polynomials in a Computer: Traditional Approach. A textbook definition of a polynomial of one variable is that it is a function of the type

$$f(x) = c_0 + c_1 \cdot x + c_2 \cdot x^2 + \ldots + c_d \cdot x^d.$$

From the viewpoint of this definition, it is natural to represent a polynomial of one variable as a corresponding sequence of coefficients $c_0, c_1, c_2 \ldots, c_d$. This is exactly how polynomials of one variable are usually represented.

Similarly, a polynomial of several variables $x_1, \ldots, x_n$ is usually defined as linear combination of monomials, i.e., expressions of the type $x_1^{d_1} \cdot \ldots \cdot x_n^{d_n}$. Thus, a natural way to represent a polynomial

$$f(x_1, \ldots, x_n) = \sum_{d_1, \ldots, d_n} c_{d_1 \ldots d_n} \cdot x_1^{d_1} \cdot \ldots \cdot x_n^{d_n}$$

is to represent it as a corresponding multi-D array of coefficients $c_{d_1 \ldots d_n}$.

Bernstein Polynomials: A Description. It has been shown that in many computational problems, it is more efficient to use an alternative representation. This alternative representation was first proposed by a mathematician Bernstein, and so polynomials represented in this form are known as *Bernstein polynomials*. For functions of one variable, Bernstein proposed to represent a function as a linear combination

$$\sum_{k=0}^{d} c_k \cdot (x - a)^k \cdot (b - x)^{d-k}$$

of special polynomials

$$p_k(x) = (x - a)^k \cdot (b - x)^{d-k}.$$

For functions of several variables, Bernstein's representation has the form

$$f(x_1, \ldots, x_n) = \sum_{k_1 \ldots k_n} c_{k_1 \ldots k_n} \cdot p_{k_1 1}(x_1) \cdot \ldots \cdot p_{k_n n}(x_n),$$

where

$$p_{k_i i}(x_i) \stackrel{\text{def}}{=} (x_i - a_i)^{k_i} \cdot (b_i - x_i)^{d-k_i}.$$

In this representation, we store the coefficients $c_{k_1 \ldots k_n}$ in the computer.

Bernstein polynomials are actively used, e.g., in computer graphics and computer-aided design, where they are not only more computationally efficient, but they also lead – within a comparable computation time – to smoother and more stable descriptions than traditional computer representations of polynomials. In many applications in which we are interested in functions defined on a given interval $[\underline{x},\overline{x}]$ – or a given multi-D box – we get better results if instead, we represent a general polynomial as a linear combination of *Bernstein polynomials*, i.e., functions proportional to

$$(x - \underline{x})^k \cdot (\overline{x} - x)^{n-k},$$

and store coefficients of this linear combination; see, e.g., [45, 46, 47, 48, 97, 137].

Example. A straightforward way to represent a quadratic polynomial $f(x) = c_0 + c_1 \cdot x + c_2 \cdot x^2$ is to store the coefficients c_0, c_1, and c_2. In the Bernstein representation, a general quadratic polynomial on the interval $[0,1]$ can be represented as

$$f(x) = a_0 \cdot x^0 \cdot (1-x)^2 + a_1 \cdot x^1 \cdot (1-x)^1 + a_0 \cdot x^2 \cdot (1-x)^0 = a_0 \cdot x^2 + a_1 \cdot x \cdot (1-x) + a_2 \cdot (1-x)^2;$$

to represent a generic polynomial in a computer, we store the values a_0, a_1, and a_2. (To be more precise, we store values proportional to a_i.)

Natural Questions. Natural questions are:

- why is the use of these basic functions more efficient than the use of standard monomials $\prod_{i=1}^{n} x_i^{k_i}$?
- are Bernstein polynomials the best or these are even better expressions?

Towards Possible Answers to These Questions. To answer these questions, we take into account that in the 1-D case, an interval $[\underline{x},\overline{x}]$ is uniquely determined by its endpoints $\underline{x}$ and $\overline{x}$. Similarly, in the multi-D case, a general box $[\underline{x}_1,\overline{x}_1] \times \ldots \times [\underline{x}_n,\overline{x}_n]$ is uniquely determined by two multi-D "endpoints" $\underline{x} = (\underline{x}_1,\ldots\underline{x}_n)$ and $\overline{x} = (\overline{x}_1,\ldots,\overline{x}_n)$. It is therefore reasonable to design the basic polynomials as follows:

- first, we find two polynomial functions $\underline{f}(x)$ and $\overline{f}(x)$, where $x = (x_1,\ldots,x_n)$, related to each of the endpoints;
- then, we use some combination operation $F(a,b)$ to combine the functions $\underline{f}(x)$ and $\overline{f}(x)$ into a single function $f(x) = G(\underline{f}(x),\overline{f}(x))$.

In this section, we use the approach from [115] to prove that if we select the optimal polynomials $\underline{f}(x)$ and $\overline{f}(x)$ on the first stage and the optimal combination operation on the second stage, then the resulting function $f(x)$ is proportional to a Bernstein polynomial. This result first appeared in [108].

In other words, we prove that under reasonable optimality criteria, Bernstein polynomials can be uniquely determined from the requirement that they are optimal combinations of optimal polynomials corresponding to the interval's endpoints.

Formulation of the Problem: Reminder. Let us first find optimal polynomials corresponding to endpoints $x^{(0)} = \underline{x}$ and $x^{(0)} = \overline{x}$.

We consider applications in which the dependence of a quantity y on the input values $x_1,\ldots,x_n$ is approximated by a polynomial $y = f(x) = f(x_1,\ldots,x_n)$. For each of the two endpoints $x^{(0)} = \underline{x}$ and $x^{(0)} = \overline{x}$, out of all polynomials which are "related" to this point, we want to find the one which is, in some reasonable sense, optimal.

How to Describe This Problem in Precise Terms. To describe this problem in precise terms, we need to describe:

- what it means for a polynomial to be "related" to the point, and
- what it means for one polynomial to be "better" than the other.

Physical Meaning. To formalize the two above notions, we take into account that in many practical applications, the inputs numbers x_i are values of some physical quantities, and the output y also represent the value of some physical quantity.

Scaling and Shift Transformations. The numerical value of each quantity depends on the choice of a measuring unit and on the choice of the starting point. If we replace the original measuring unit by a unit which is λ times smaller (e.g., use centimeters instead of meters), then instead of the original numerical value y, we get a new value $y' = \lambda \cdot y$.

Similarly, if we replace the original starting point with a new point which corresponds to y_0 on the original scale (e.g., as the French Revolution did, select 1789 as the new Year 0), then, instead as the original numerical value y, we get a new numerical value $y' = y - y_0$.

In general, if we change both the measuring unit and the starting point, then instead of the original numerical value y, we get the new value $\lambda \cdot y - y_0$.

We Should Select a Family of Polynomials. Because of scaling and shift, for each polynomial $f(x)$, the polynomials $\lambda \cdot f(x) - y_0$ represent the same dependence, but expressed in different units. Because of this fact, we should not select a *single* polynomial, we should select the entire *family* $\{\lambda \cdot f(x) - y_0\}_{\lambda, y_0}$ of polynomials representing the original dependence for different selections of the measuring unit and the starting point.

Scaling and Shift for Input Variables. In many practical applications, the inputs numbers x_i are values of some physical quantities. The numerical value of each such quantity also depends on the choice of a measuring unit and on the choice of the starting point. By using different choices, we get new values $x'_i = \lambda_i \cdot x_i - x_{i0}$, for some values λ_i and x_{i0}.

Transformations Corresponding to a Given Endpoint $x^{(0)} = \left(x_1^{(0)}, \ldots, x_n^{(0)} \right)$. Once the endpoint is given, we no longer have the freedom of changing the starting point, but we still have re-scalings: $x_i - x_i^{(0)} \rightarrow \lambda_i \cdot \left(x_i - x_i^{(0)} \right)$, i.e., equivalently,

$$x_i \rightarrow x'_i = x_i^{(0)} + \lambda \cdot \left(x_i - x_i^{(0)} \right).$$

What Is Meant by "The Best" Family? When we say "the best" family, we mean that on the set of all the families, there is a relation $\succeq$ describing which family is better or equal in quality. This relation must be transitive (if $\mathscr{F}$ is better than $\mathscr{G}$, and $\mathscr{G}$ is better than $\mathscr{H}$, then $\mathscr{F}$ is better than $\mathscr{H}$).

Final Optimality Criteria. The preference relation $\succeq$ is not necessarily asymmetric, because we can have two families of the same quality. However, we would like to require that this relation be *final* in the sense that it should define a unique *best* family $\mathscr{F}_{opt}$, for which $\forall \mathscr{G} \, (\mathscr{F}_{opt} \succeq \mathscr{G})$.

Indeed, if none of the families is the best, then this criterion is of no use, so there should be *at least one* optimal family.

If *several* different families are equally best, then we can use this ambiguity to optimize something else: e.g., if we have two families with the same approximating quality, then we choose the one which is easier to compute. As a result, the original criterion was not final: we obtain a new criterion: $\mathscr{F} \succeq_{new} \mathscr{G}$, if either $\mathscr{F}$ gives a better approximation, or if $\mathscr{F} \sim_{old} \mathscr{G}$ and $\mathscr{G}$ is easier to compute. For the new optimality criterion, the class of optimal families is narrower.

We can repeat this procedure until we obtain a final criterion for which there is only one optimal family.

Optimality Criteria Should Be Invariant. Which of the two families is better should not depend on the choice of measuring units for measuring the inputs x_i. Thus, if $\mathscr{F}$ was better than $\mathscr{G}$, then after re-scaling, the re-scaled family $\mathscr{F}$ should still be better than the re-scaled family $\mathscr{G}$.

Thus, we arrive at the following definitions.

Definition 3.5.1. *By a family, we mean a set of functions from $\mathbb{R}^n \to \mathbb{R}$ which has the form $\{C \cdot f(x) - y_0 : C, y_0 \in \mathbb{R}, C > 0\}$ for some polynomial $f(x)$. Let $\mathscr{F}$ denote the class of all possible families.*

Definition 3.5.2. *By a optimality criterion $\preceq$ on the class $\mathscr{F}$, we mean a pre-ordering relation on the set $\mathscr{F}$, i.e., a transitive relation for which $F \preceq F$ for every F. We say that a family F is optimal with respect to the optimality criterion $\preceq$ if $G \preceq F$ for all $G \in \mathscr{F}$.*

Definition 3.5.3. *We say that the optimality criterion is final if there exists one and only one optimal family.*

Definition 3.5.4. *Let $x^{(0)}$ be a vector. By a $x^{(0)}$-rescaling corresponding to the values $\lambda = (\lambda_1, \dots, \lambda_n)$, $\lambda_i > 0$, we mean a transformation $x \to x' = T_{x^{(0)}, \lambda}(x)$ for which*

$$x'_i = x_i^{(0)} + \lambda_i \cdot \left(x_i - x_i^{(0)}\right).$$

By a $x^{(0)}$-rescaling of a family $F = \{C \cdot f(x) - y_0\}_{C,y_0}$, we mean a family $T_{x^{(0)}, \lambda}(F) = \{C \cdot f(T_{x^{(0)}, \lambda}(x))s\}_{C,y_0}$. We say that an optimality criterion is $x^{(0)}$-scaling-invariant if for every F, G, and λ, $F \preceq G$ implies $T_{x^{(0)}, \lambda}(F) \preceq T_{x^{(0)}, \lambda}(G)$.

Proposition 3.5.1. *Let $\preceq$ be a final $x^{(0)}$-scaling-invariant optimality criterion. Then every polynomial from the optimal family has the form*

$$f(x) = A + B \cdot \prod_{i=1}^{n} \left(x_i - x_i^{(0)}\right)^{k_i}.$$

Comment. For readers' convenience, all the proofs are placed in the special (last) Proofs section.

Discussion. As we have mentioned, the value of each quantity is defined modulo a starting point. It is therefore reasonable, for y, to select a starting point so that $A = 0$. Thus, we get the dependence

$$f(x) = B \cdot \prod_{i=1}^{n} \left(x_i - x_i^{(0)} \right)^{k_i}.$$

Once the starting point for y is fixed, the only remaining y-transformations are scalings $y \to \lambda \cdot y$.

Optimal Combination Operations. In the previous section, we described the optimal functions corresponding to the endpoints $\underline{x}$ and $\bar{x}$. What is the optimal way of combining these functions? Since we are dealing only with polynomial functions, it is reasonable to require that a combination operation transform polynomials into polynomials.

Definition 3.5.5. *By a* combination operation, *we mean a function* $K : \mathbb{R}^2 \to \mathbb{R}$ *for which, if* $\underline{f}(x)$ *and* $\bar{f}(x)$ *are polynomials, then the composition* $K\left(\underline{f}(x), \bar{f}(x)\right)$ *is also a polynomial.*

Lemma 3.5.1. *A function* $K(a,b)$ *is a combination operation if and only if it is a polynomial.*

Discussion. Similarly to the case of optimal functions corresponding to individual endpoint, the numerical value of the function $K\left(\underline{a}, \bar{a}\right)$ depends on the choice of the measuring unit and the starting point: an operation that has the form $K\left(\underline{a}, \bar{a}\right)$ under one choice of the measuring unit and starting point has the form $C \cdot K\left(\underline{a}, \bar{a}\right) - y_0$ under a different choice. Thus, we arrived at the following definition.

Definition 3.5.6. *By a* C-family, *we mean a set of functions from* $\mathbb{R}^2 \to \mathbb{R}$ *which has the form* $\{C \cdot K(a,b) - y_0 : C, y_0 \in \mathbb{R}, C > 0\}$ *for some combination operation* $K(a,b)$. *Let* $\mathcal{K}$ *denote the class of all possible C-families.*

Definition 3.5.7. *By an* optimality criterion $\preceq$ *on the class* $\mathcal{K}$ *of all C-families, we mean a pre-ordering relation on the set* $\mathcal{K}$, *i.e., a transitive relation for which* $F \preceq F$ *for every C-family F. We say that a C-family F is* optimal *with respect to the optimality criterion* $\preceq$ *if* $G \preceq F$ *for all* $G \in \mathcal{K}$.

Definition 3.5.8. *We say that the optimality criterion is* final *if there exists one and only one optimal C-family.*

Discussion. From the previous section, we know that both functions $\underline{f}(x)$ and $\bar{f}(x)$ are determined modulo scaling $\underline{f}(x) \to \underline{\lambda} \cdot \underline{f}(x)$ and $\bar{f}(x) \to \bar{\lambda} \cdot \bar{f}(x)$. Thus, it is reasonable to require that the optimality relation not change under such re-scalings.

Definition 3.5.9. *By a* C-rescaling *corresponding to the values* $\lambda = \left(\underline{\lambda}, \bar{\lambda} \right)$, *we mean a transformation* $T_{\lambda}\left(\underline{a}, \bar{a} \right) = \left(\underline{\lambda} \cdot \underline{a}, \bar{\lambda} \cdot \bar{a} \right)$. *By a* C-rescaling *of a family*

$$F = \{C \cdot K(\underline{a}, \overline{a}) - y_0\}_{C, y_0},$$

we mean a family $T_\lambda(F) = \{C \cdot K(T_\lambda(a))\}_{C, y_0}$. We say that an optimality criterion is C-scaling-invariant if for every F, G, and λ, $F \preceq G$ implies $T_\lambda(F) \preceq T_\lambda(G)$.

Proposition 3.5.2. Let $\preceq$ be a final C-scaling-invariant optimality criterion. Then every combination operation from the optimal family has the form

$$K(\underline{a}, \overline{a}) = A + B \cdot \underline{a}^k \cdot \overline{a}^{\overline{k}}.$$

Conclusions. By applying this optimal combination operation from Section 3.5.4 to the optimal functions corresponding to $x^{(0)} = \underline{x}$ and $x^{(0)} = \overline{x}$ (described in Section 3.5.3), we conclude that the resulting function has the form

$$f(x_1, \ldots, x_n) = K\left(\underline{f}(x_1, \ldots, x_n), \overline{f}(x_1, \ldots, x_n)\right) =$$

$$A + B \cdot \left(\prod_{i=1}^{n} (x_i - \underline{x}_i)^{\underline{k}_i}\right)^{\underline{k}} \cdot \left(\prod_{i=1}^{n} (\overline{x}_i - x_i)^{\overline{k}_i}\right)^{\overline{k}}.$$

Modulo an additive constant, this function has the form

$$f(x_1, \ldots, x_n) = B \cdot \prod_{i=1}^{n} (x_i - \underline{x}_i)^{\underline{k}_i'} \cdot \prod_{i=1}^{n} (\overline{x}_i - x_i)^{\overline{k}_i'},$$

where $\underline{k}_i' = \underline{k}_i \cdot \underline{k}$ and $\overline{k}_i' = \overline{k}_i \cdot \overline{k}$.

These are Bernstein polynomials. Thus, Bernstein polynomials can indeed by uniquely determined as the result of applying an optimal combination operation to optimal functions corresponding to $\underline{x}$ and $\overline{x}$.

Proofs

Proof of Proposition 3.5.1

$1°$. Let us first prove that the optimal family F_{opt} is $x^{(0)}$-scaling-invariant, i.e., $T_{x^{(0)}, \lambda}(F_{\mathrm{opt}}) = F_{\mathrm{opt}}$.

Since F_{opt} is an optimal family, we have $G \preceq F_{\mathrm{opt}}$ for all families G. In particular, for every family G and for every λ, we have $T_{x^{(0)}, \lambda^{-1}}(G) \preceq F_{\mathrm{opt}}$. Since the optimal criterion is $x^{(0)}$-scaling-invariant, we conclude that

$$T_{x^{(0)}, \lambda}\left(T_{x^{(0)}, \lambda^{-1}}(G)\right) \preceq T_{x^{(0)}, \lambda}(F_{\mathrm{opt}}).$$

One can easily check that if we first re-scale the family with the coefficient λ^{-1}, and then with λ, then we get the original family G back. Thus, the above conclusion takes the form $G \preceq T_{x^{(0)}, \lambda}(F_{\mathrm{opt}})$. This is true for all families G, hence the family $T_{x^{(0)}, \lambda}(F_{\mathrm{opt}})$ is optimal. Since the optimality criterion is final, there is only one optimal family, so $T_{x^{(0)}, \lambda}(F_{\mathrm{opt}}) = F_{\mathrm{opt}}$. The statement is proven.

$2°$. For simplicity, instead of the original variables x_i, let us consider auxiliary variables $z_i = x_i - x_i^{(0)}$. In terms of these variables, re-scaling takes a simpler form $z_i \to \lambda_i \cdot z_i$. Since $x_i = z_i + x_i^{(0)}$, the dependence $f(x_1, \ldots, x_n)$ take the form

$$g(z_1, \ldots, z_n) = f\left(z_1 + x_1^{(0)}, \ldots, z_n + x_n^{(0)}\right).$$

Since the function $f(x_1, \ldots, x_n)$ is a polynomial, the new function $g(z_1, \ldots, z_n)$ is a polynomial too.

$3°$. Let us now use the invariance that we have proved in Part 1 of this proof to find the dependence of the function $f(z)$ on each variable z_i. For that, we use invariance under transformations that change z_i to $\lambda_i \cdot z_i$ and leave all other coordinates z_j ($j \neq i$) intact.

Let us fix the values z_j of all the variables except for z_i. Under the above transformation, invariance implies that if $g(z_1, \ldots, z_{i-1}, z_i, z_{i+1}, \ldots, z_n)$ is a function from the optimal family, then the re-scaled function $g(z_1, \ldots, z_{i-1}, \lambda_i \cdot z_i, z_{i+1}, \ldots, z_n)$ belongs to the same family, i.e.,

$$g(z_1, \ldots, z_{i-1}, \lambda_i \cdot z_i, z_{i+1}, \ldots, z_n) = C(\lambda_i) \cdot g(z_1, \ldots, z_{i-1}, z_i, z_{i+1}, \ldots, z_n) - y_0(\lambda_i)$$

for some values C and y_0 depending on λ_i. Let us denote

$$g_i(z_i) = g(z_1, \ldots, z_{i-1}, z_i, z_{i+1}, \ldots, z_n).$$

Then, the above condition takes the form

$$g_i(\lambda \cdot z_i) = C(\lambda_i) \cdot g_i(z_i) - y_0(\lambda_i).$$

It is possible that the function $g_i(z_i)$ is a constant. If it is not a constant, this means that there exist values $z_i \neq z_i'$ for which $g_i(z_i) \neq g_i(z_i')$. For these two values, we get

$$g_i(\lambda_i \cdot z_i) = C(\lambda_i) \cdot g_i(z_i) - y_0(\lambda_i);$$

$$g_i(\lambda_i \cdot z_i') = C(\lambda_i) \cdot g_i(z_i') - y_0(\lambda_i).$$

By subtracting these equations, we conclude that

$$g_i(\lambda_i \cdot z_i) - g_i(\lambda_i \cdot z_i') = C(\lambda_i) \cdot (g_i(z_i) - g_i(z_i')),$$

hence

$$C(\lambda_i) = \frac{g_i(\lambda_i \cdot z_i) - g_i(\lambda_i \cdot z_i')}{g_i(z_i) - g_i(z_i')}.$$

Since the function $g_i(z_i)$ is a polynomial, the right-hand side is a smooth function of λ. Thus, the dependence of $C(\lambda_i)$ on λ_i is differentiable (smooth). Since

$$y_0(\lambda_i) = C(\lambda_i) \cdot g_i(z_i) - g_i(\lambda_i \cdot z_i),$$

and both C and g_i are smooth functions, the dependence $y_0(\lambda_i)$ is also smooth.

Since all three functions C, y_0, and g_i are differentiable, we can differentiate both sides of the equality $g_i(\lambda_i \cdot z_i) = C(\lambda_i) \cdot g_i(z_i) - y_0(\lambda_i)$ by λ_i and take $\lambda_i = 1$. This leads to the formula

$$z_i \cdot \frac{dg_i}{dz_i} = C_1 \cdot g_i(z_i) - y_1,$$

where we denoted $C_1 \overset{\text{def}}{=} \dfrac{dC}{d\lambda_i}_{|\lambda_i=1}$ and $y_1 \overset{\text{def}}{=} \dfrac{dy_0}{d\lambda_i}_{|\lambda_i=1}$.

By moving all the terms related to g_i to one side and all the terms related to z_i to the other side, we get

$$\frac{dg_i}{C_1 \cdot g_i - y_1} = \frac{dz_i}{z_i}.$$

We consider two possibilities: $C_1 = 0$ and $C_1 \neq 0$.

3.1°. If $C_1 = 0$, then the above equation takes the form

$$-\frac{1}{y_1} \cdot dg_i = \frac{dz_i}{z_i}.$$

Integrating both sides, we get

$$-\frac{1}{y_1} \cdot g_i = \ln(z_i) + \text{const},$$

thus $g_i = -y_1 \cdot \ln(z_i) + \text{const}$. This contradicts to the fact that the dependence $g_i(z_i)$ is polynomial. Thus, $C_1 \neq 0$.

3.2°. Since $C_1 \neq 0$, we can introduce a new variable $h_i = g_i - \dfrac{y_1}{C_1}$. For this new variable, we have $dh_i = dg_i$. Hence the above differential equation takes the simplified form

$$\frac{1}{C_1} \cdot \frac{dh_i}{h_i} = \frac{dz_i}{z_i}.$$

Integrating both sides, we get

$$\frac{1}{C_1} \cdot \ln(h_i) = \ln(z_i) + \text{const},$$

hence

$$\ln(h_i) = C_1 \cdot \ln(z_i) + \text{const},$$

and

$$h_i = \text{const} \cdot z_i^{C_1}.$$

Thus,

$$g_i(z_i) = h_i(z_i) + \frac{y_1}{C_1} = \text{const} \cdot z_i^{C_1} + \frac{y_1}{C_1}.$$

Since we know that $g_i(z_i)$ is a polynomial, the power C_1 should be a non-negative integer, so we conclude that

$$g_i(z_i) = A \cdot z_i^{k_i} + B$$

for some values A_i, B_i, and k_i which, on general, depend on all the other values z_j.

4°. Since the function $g(z_1, \ldots, z_n)$ is a polynomial, it is continuous and thus, the value k_i continuously depends on z_j. Since the value k_i is always an integer, it must therefore be constant – otherwise we would have a discontinuous jump from one integer to another. Thus, the integer k_i is the same for all the values z_j.

5°. Let us now use the above dependence on each variable z_i to find the dependence on two variables. Without losing generality, let us consider dependence on the variables z_1 and z_2.

Let us fix the values of all the other variables except for z_1 and z_2, and let us define

$$g_{12}(z_1, z_2) = g(z_1, z_2, z_3, \ldots, z_n).$$

Our general result can be applied both to the dependence on z_1 and to the dependence on z_2. The z_1-dependence means that $g_{12}(z_1, z_2) = A_1(z_2) \cdot z_1^{k_1} + B_1(z_2)$, and the z_1-dependence means that $g_{12}(z_1, z_2) = A_2(z_1) \cdot z_2^{k_2} + B_2(z_1)$. Let us consider two possible cases: $k_1 = 0$ and $k_1 \neq 0$.

5.1°. If $k_1 = 0$, this means that $g_{12}(z_1, z_2)$ does not depend on z_1 at all, so both A_2 and B_2 do not depend on z_1, hence we have $g_{12}(z_1, z_1) = A_2 \cdot z_2^{k_2} + B_2$.

5.2°. Let us now consider the case where $k_1 \neq 0$. For $z_1 = 0$, the z_1-dependence means that $g_{12}(0, z_2) = B_1(z_2)$, and the z_2-dependence implies that $B_1(z_2) = g_{12}(0, z_2) = A_2(0) \cdot z_2^{k_2} + B_2(0)$.

For $z_1 = 1$, the z_1-dependence means that $g_{12}(1, z_2) = A_1(z_2) + B_1(z_2)$. On the other hand, from the z_2-dependence, we conclude that $A_1(z_2) + B_1(z_2) = g_{12}(1, z_2) = A_2(1) \cdot z_2^{k_2} + B_2(1)$. We already know the expression for $B_1(z_2)$, so we conclude that

$$A_1(z_2) = g_{12}(1, z_2) - B_1(z_2) = (A_2(1) - A_2(0)) \cdot z_2^{k_2} + (B_2(1) - B_2(0)).$$

Thus, both $A_1(z_2)$ and $B_1(z_2)$ have the form $a + b \cdot z_2^{k_2}$, hence we conclude that

$$g_{12}(z_1, z_2) = (a + b \cdot z_2^{k_2}) \cdot z_1^{k_1} + (c + d \cdot z_2^{k_2}) = c + a \cdot z_1^{k_1} + d \cdot z_2^{k_2} + b \cdot z_1^{k_1} \cdot z_2^{k_2}.$$

Previously, we only considered transformations of a single variable, let us now consider a joint transformation $z_1 \to \lambda_1 \cdot z_1$, $z_2 \to \lambda_2 \cdot z_2$. In this case, we get

$$g(\lambda_1 \cdot z_1, \lambda_2 \cdot z_2) = c + a \cdot \lambda_1^{k_1} \cdot z_1^{k_1} + d \cdot \lambda_2^{k_2} \cdot z_2^{k_2} + b \cdot \lambda_1^{k_1} \cdot \lambda_2^{k_2} \cdot z_1^{k_1} \cdot z_2^{k_2}.$$

We want to make sure that

$$g(\lambda_1 \cdot z_1, \lambda_2 \cdot z_2) = C(\lambda_1, \lambda_2) \cdot g(z_1, z_2) - y_0(\lambda_1, \lambda_2),$$

i.e., that

$$c + a \cdot \lambda_1^{k_1} \cdot z_1^{k_1} + d \cdot \lambda_2^{k_2} \cdot z_2^{k_2} + b \cdot \lambda_1^{k_1} \cdot \lambda_2^{k_2} \cdot z_1^{k_1} \cdot z_2^{k_2} =$$

$$C(\lambda_1, \lambda_2) \cdot (c + a \cdot z_1^{k_1} + d \cdot z_2^{k_2} + b \cdot z_1^{k_1} \cdot z_2^{k_2}) - y_0(\lambda_1, \lambda_2).$$

Both sides are polynomials in z_1 and z_2; the polynomials coincide for all possible values z_1 and z_2 if and only if all their coefficients coincide. Thus, we conclude that

$$a \cdot \lambda_1^{k_1} = a \cdot C(\lambda_1, \lambda_2);$$

$$d \cdot \lambda_2^{k_2} = d \cdot C(\lambda_1, \lambda_2);$$

$$c \cdot \lambda_1^{k_1} \cdot \lambda_2^{k_2} = c \cdot C(\lambda_1, \lambda_2).$$

If $a \neq 0$, then by dividing both sides of the a-containing equality by a, we get $C(\lambda_1, \lambda_2) = \lambda_1^{k_1}$. If $d \neq 0$, then by dividing both sides of the d-containing equality by d, we get $C(\lambda_1, \lambda_2) = \lambda_2^{k_2}$. If $c \neq 0$, then by dividing both sides of the c-containing equality by c, we get $C(\lambda_1, \lambda_2) = \lambda_1^{k_1} \cdot \lambda_2^{k_2}$. These three formulas are incompatible, so only one of three coefficients a, d, and c is different from 0 and two other coefficients are equal to 0. In all three cases, the dependence has the form

$$g_{12}(z_1, z_2) = a + \text{const} \cdot z_1^{\ell_1} \cdot z_2^{\ell_2}.$$

$6°$. Similarly, by considering more variables, we conclude that

$$g(z_1, \ldots, z_n) = a + \text{const} \cdot z_1^{\ell_1} \cdot \ldots \cdot z_n^{\ell_n}.$$

By plugging in the values z_i in terms of x_i, we get the conclusion of the proposition. The proposition is proven.

Proof of Lemma 3.5.1. Let us first show that if the function $K(a, b)$ is a combination operation, then $K(a, b)$ is a polynomial. Indeed, by definition of a combination operation, if we take $\underline{f}(x) = x_1$ and $\overline{f}(x) = x_2$, then the function $f(x) = K\left(\underline{f}(x), \overline{f}(x)\right) = K(x_1, x_2)$ is a polynomial.

Vice versa, if $K(x_1, x_2)$ is a polynomial, then for every two polynomials $\underline{f}(x)$ and $\overline{f}(x)$, the composition $f(x) = K\left(\underline{f}(x), \overline{f}(x)\right)$ is also a polynomial. The lemma is proven.

Proof of Proposition 3.5.2. Due to Lemma, Proposition 3.5.2 follows from Proposition 3.5.1 – for the case of two variables.

Need for an Intuitive Explanation. In the above text, we provided a mathematically complex symmetry-based explanation of why Bernstein polynomials work well. This explanation is far from being intuitively clear, so it is desirable to supplement this mathematical result with an intuitive explanation.

Preliminary Step: Reducing All Intervals to the Interval $[0, 1]$**.** To provide the desired intuitive explanation, we use *fuzzy logic*– a technique for describing informal intuitive arguments. This explanation first appeared in [103]. We want to use fuzzy logic to analyze polynomial and piece-wise polynomial approximations. In fuzzy logic, traditionally, possible truth values form an interval $[0, 1]$. In some intelligent systems, other intervals are used – e.g., in the historically first expert system MYCIN the interval $[-1, 1]$ was used to describe possible degrees of confidence. It is well known that it does not matter much what interval we use since we can easily reduce values x from an interval $[a, b]$ to values t from the interval $[0, 1]$ by taking $t = \dfrac{x - a}{b - a}$; vice versa, once we know the new value t, we can easily reconstruct the original value x as $x = a + t \cdot (b - a)$.

To facilitate the use of traditional fuzzy techniques, let us therefore reduce all the intervals $[a_i, b_i]$ to the interval $[0, 1]$. In other words, instead of the original function

$$f(x_1, \ldots, x_n) : [a_1, b_1] \times \ldots \times [a_n, b_n] \to \mathbb{R},$$

we consider a new function

$$F(t_1, \ldots, t_n) : [0, 1]^n \to \mathbb{R},$$

which is defined as

$$F(t_1, \ldots, t_n) = f(a_1 + t_1 \cdot (b_1 - a_1), \ldots, a_n + t_n \cdot (b_n - a_n)).$$

Vice versa, if we find a good approximation $\widetilde{F}(t_1, \ldots, t_n)$ to the new function $F(t_1, \ldots, t_n)$, we can easily generate an approximation $\widetilde{f}(x_1, \ldots, x_n)$ to the original function $f(x_1, \ldots, x_n)$ as follows:

$$\widetilde{f}(x_1, \ldots, x_n) = \widetilde{F}\left(\frac{x_1 - a_1}{b_1 - a_1}, \ldots, \frac{x_n - a_n}{b_n - a_n}\right).$$

Fuzzy-Based Function Approximations: Reminder. Fuzzy techniques have been actively used to approximate functional dependencies: namely, such dependencies are approximated by fuzzy rules; see, e.g., [70, 73, 120]. The simplest case is where each rule has a fuzzy condition and a crisp conclusion, i.e., has the type

"if x is P, then $y = c$",

where P is a fuzzy property (such as "small") characterized by a membership function $\mu(x)$, and c is a real number. For the case of several inputs, we have rules of the type

"if x_1 is P_1, x_2 is P_2, ..., and x_n is P_n, then $y = c$."

The degree to which a given input x_i satisfies the property P_i is equal to $\mu_i(x_i)$, where $\mu_i(x)$ is the membership function corresponding to the property P_i. The degree to which the tuple $(x_1, \ldots, x_n)$ satisfies the condition of the rule – i.e., the statement

"x_1 is P_1, x_2 is P_2, ..., and x_n is P_n"

– is therefore equal to $f_\&(\mu_1(x_1), \ldots, \mu_n(x_n))$, where $f_\&$ is an appropriate t-norm ("and"-operation). One of the simplest t-norms is the product $f_\&(a, b) = a \cdot b$. For this t-norm, the degree d to which the above rule is satisfied is equal to the product $\mu_1(x_1) \cdot \ldots \cdot \mu_n(x_n)$ of the corresponding membership degrees.

When we have several rules, then we get different conclusions $c_1, \ldots, c_r$ with degrees $d_1, \ldots, d_r$; we need to come form a single value that combines these conclusions. The larger the degree d_i, the more weight we should give to the conclusion c_i. A natural way is thus simply to take the weighted average $c_1 \cdot d_1 + \ldots + c_r \cdot d_r$. This weighted average can be interpreted in fuzzy terms if we interpret the combination as the following statement:

- "either (the condition for the 1st rule is satisfied and its conclusion is satisfied)
- or (the condition for the 2nd rule is satisfied and its conclusion is satisfied)
- or . . .
- or (the condition for the r-th rule is satisfied and its conclusion is satisfied),"

where we describe "and" as multiplication and "or" as addition.

Resulting Interpretation of the Usual Polynomial Representation. The functions of one variable, the traditional computer representations of a polynomial has the form

$$c_0 + c_1 \cdot x + c_2 \cdot x^2 + \ldots + c_m \cdot x^m.$$

The corresponding approximation can be interpreted as the following set of fuzzy rules:

- c_0 (with no condition);
- if x, then c_1;
- if x^2, then c_2; . . .
- if x^m, then c_m.

In fuzzy logic, if we take x as the degree to which $x \in [0, 1]$ is large, then:

- x^2 is usually interpreted as "very large",
- $x^4 = (x^2)^2$ is interpreted as "very very large",
- $x^8 = (x^4)^2 = ((x^2)^2)^2$ is interpreted as "very very very large", etc., and
- intermediate powers x^3, x^5, x^7, etc., are interpreted as as some intermediate hedges.

Thus, the above rules have the form:

- c_0;
- if x is large, then c_1;
- if x is very large, then c_2, etc.

Similarly, for polynomials of several variables, we have as many rules as there are monomials $c_{k_1 \ldots k_n} \cdot x_1^{k_1} \cdot \ldots \cdot x_n^{k_n}$. For example, a monomial

$$c_{012} \cdot x_1^0 \cdot x_1^1 \cdot x_2^2 = c_{012} \cdot x_2 \cdot x_3^2$$

corresponds to the following rule:

"if x_2 is large and x_3 is very large, then c_{012}."

Fuzzy Interpretation Reveals Limitations of the Traditional Computer Representation of Polynomials. From the fuzzy viewpoint, there are two limitations to this interpretation.

The first limitation is related to the fact that an accurate representation requires polynomials of higher degrees, with several distinct coefficients corresponding to different hedges such as "very", "very very", etc. In practice, we humans can only meaningfully distinguish between a small number of hedges, and this limits the possibility of meaningfully obtaining such rules from experts.

The second limitation is that for the purposes of computational efficiency, it is desirable to have a computer representation in which as few terms as possible are needed to represent each function. This can be achieved if in some important cases, some of the coefficients in the corresponding computer representation are close to 0 and can, therefore, be safely ignored. For the above fuzzy representation, all the terms are meaningful, and there seems to be no reason why some of these terms can be ignored.

How Can We Overcome Limitations of the Traditional Computer Representation: Fuzzy Analysis of the Problem. From the fuzzy viewpoint, the traditional computer representation of polynomials corresponds to taking into account the opinions of a single expert. Theoretically, we can achieve high accuracy this way if we have an expert who can meaningfully distinguish between "large", "very large", "very very large", etc. However, most experts are not very good in such a distinction. A typical expert is at his or her best when this expert distinguishes between "large" and "not large", any more complex distinctions are much harder.

Since we cannot get a good approximation by using a *single* expert, why not use *multiple* experts? In this case, there is no need to force an expert into making a difficult distinction between "very large" and "very very large". So, we can as well use each expert where each expert is the strongest: by requiring each expert to distinguish between "large" and "very large". In this setting, once we have d experts, for each variable x_i, we have the following options:

- The first option is where all d experts believe that x_i is large: the 1st expert believes that x is large, the 2nd believes that x_i is large, etc. Since we have decided to use product for representing "and", the degree to which this condition is satisfied is equal to $x_i \cdot \ldots \cdot x_i = x_i^d$.
- Another option is where $d - 1$ experts believe that x_i is large, and the remaining expert believes that x is not large. The corresponding degree is equal to $x_i^{d-1} \cdot (1 - x_i)$.
- In general, we can have k_i experts believing that believing that x_i is large and $d - k_i$ experts believing that x is not large. The corresponding degree is equal to

$$x_i^{k_i} \cdot (1 - x_i)^{d - k_i}.$$

For this variable, general weighted combinations of such rules lead to polynomials of the type $\sum_{k_i} c_{k_i} \cdot x_i^{k_i} \cdot (1 - x_i)^{d - k_i}$, i.e., to Bernstein polynomials of one variable.

For several variables, we have the degree $p_{k_i i}(x_i) = x_i^{k_i} \cdot (1 - x_i)^{d - k_i}$ with which each variable x_i satisfies the corresponding condition. Hence, the degree to which all n variables satisfy the corresponding condition is equal to the product $p_{k_1 1}(x_1) \cdot \ldots \cdot p_{k_n n}(x_n)$ of these degrees. Thus, the corresponding fuzzy rules lead to polynomials of the type

$$\sum_{k_1, \ldots, k_n} c_{k_1 \ldots k_n} \cdot p_{k_1 1}(x_1) \cdot \ldots \cdot p_{k_n n}(x_n),$$

i.e., to *Bernstein polynomials*.

So, we indeed get a fuzzy explanations for the emergence of Bernstein polynomials.

Why Bernstein Polynomials Are More Computationally Efficient: A Fuzzy Explanation. Let us show that the above explanation of the Bernstein polynomials leads to the desired explanation of why the Bernstein polynomials are more computationally efficient than the traditional computer representation of polynomials.

Indeed, the traditional polynomials correspond to rules in which conditions are "x is large", "x is very large", "x is very very large", etc. It may be difficult to distinguish between these terms, but there is no reason to conclude that some of the corresponding terms become small.

In contrast, each term $x_i^{k_i} \cdot (1 - x_i)^{d - k_i}$ from a Bernstein polynomial, with the only exception of cases $k_i = 0$ and $k_i = D$, corresponds to the condition of the type

- "x_i is large (very large, etc.) *and*
- x_i is not large (very not large, etc.)".

While in fuzzy logic, such a combination is possible, there are important cases where this value is close to 0 – namely, in the practically important cases where we are either confident that x_i is large or we are confident that x_i is not large. In these cases, the corresponding terms can be safely ignored, and thus, computations become indeed more efficient.

From Fuzzy Explanation to a More Precise Explanation: Main Idea. Let us show that terms $x_i^{k_i} \cdot (1 - x_i)^{d - k_i}$ corresponding to $k_i \in (0, d)$ are indeed smaller and thus, some of them can indeed be safely ignored. To prove this fact, let us pick a threshold $\varepsilon > 0$ ($\varepsilon \ll 1$) and in each computer representation of polynomials, let us only keep the terms for which the largest possible value of this term does not exceed ε.

Traditional Computer Representation of Polynomials: Analysis. In the traditional representation, the terms are of the type $x_1^{k_1} \cdot \ldots \cdot x_n^{k_n}$. When $x_i \in [0, 1]$, each such term is a product of the corresponding terms $x_i^{k_i}$. The resulting non-negative function is increasing in all its variables, and thus, its largest possible value is attained when all the variables x_i attain their largest possible value 1. The corresponding largest value is equal to $1^{k_1} \cdot \ldots \cdot 1^{k_n} = 1$.

Since the largest value of each term is 1, and 1 is larger than the threshold ε, all the terms are retained. If we restrict ourselves to terms of order $\leq d$ for each variable x_i, we get:

- $d+1$ possible terms for one variable:

$$x_i^0 = 1, \ x_i^1 = x, \ x_i^2, \ \ldots, \ x_i^d,$$

- $(n+1)^2$ terms $x_1^{k_1} \cdot x_2^{k_2}$ for two variables,
- ..., and
- $(d+1)^n$ terms in the general case of n variables.

This number of retained terms grows exponentially with the number of variables n.

Bernstein Polynomials: Analysis. For Bernstein polynomials, each term has the product form $p_{k_1 1}(x_1) \cdot \ldots \cdot p_{k_n n}(x_n)$, where $p_{k_i i}(x_i) = x_i^{k_i} \cdot (1-x_i)^{d-k_i}$. The product of non-negative numbers $p_{k_i i}(x_i)$ is a monotonic function of its factors. Thus, its maximum is attained when each of the factors $p_{k_i i}(x_i) = x_i^{k_i} \cdot (1-x_i)^{d-k_i}$ is the largest possible. Differentiating this expression with respect to x_i, taking into account that the derivative of $f(x) = x^k$ is equal to $\dfrac{k}{x} \cdot f(x)$, and equating the resulting derivative to 0, we conclude that

$$\frac{k_i}{x_i} \cdot p_{k_i i}(x_i) - \frac{d-k_i}{1-x_i} \cdot p_{k_i i}(x_i) = 0,$$

i.e., that $\dfrac{k_i}{x_i} = \dfrac{d-k_i}{1-x_i}$. Multiplying both sides of this equality by the common denominator of the two fractions, we get

$$k_i \cdot (1-x_i) = (d-k_i) \cdot x_i,$$

i.e., $k_i - k_i \cdot x_i = d \cdot x_i - k_i \cdot x_i$. Adding $k_i \cdot x_i$ to both sides of this equation, we get $k_i = d \cdot x_i$ hence $x_i = \dfrac{k_i}{d}$. Thus, the largest value of this term is equal to

$$x_i^{k_i} \cdot (1-x_i)^{d-k_i} = \left(\frac{k_i}{d}\right)^{k_i} \cdot \left(1 - \frac{k_i}{d}\right)^{d-k_i}.$$

This value is the largest for $k_i = 0$ and $k_i = d$, when the corresponding maximum is equal to 1; as a function of k_i, it first decreases and then increases again. So, if we want to consider values for which this term is large enough, we have to consider value k_i which are close to 0 (i.e., $k_i \ll d$) or close to d (i.e., $d - k_i \ll d$).

For values k_i which are close to 0, we have $\left(1 - \dfrac{k_i}{d}\right)^{d-k_i} \approx \left(1 - \dfrac{k_i}{d}\right)^d$. It is known that for large d, this value is asymptotically equal to $\exp(-k_i)$. Thus, the logarithm of the corresponding maximum $\left(\dfrac{k_i}{d}\right)^{k_i} \cdot \left(1 - \dfrac{k_i}{d}\right)^{d-k_i}$ is asymptotically equal to the logarithm of $\left(\dfrac{k_i}{d}\right)^{k_i} \cdot \exp(-k_i)$, i.e., to $-k_i \cdot (\ln(d) - \ln(k_i) + 1)$. Since we have $k_i \ll d$, we get $\ln(k_i) \ll \ln(d)$ and therefore, the desired logarithm is asymptotically equal to $-k_i \cdot \ln(d)$.

For the values $k_i \approx d$, we can get a similar asymptotic expression $-(d - k_i) \cdot \ln(d)$. Both expressions can be described as $-\Delta_i \cdot \ln(d)$, where Δ_i denotes $\min(k_i, d - k_i)$, i.e.,

- $\Delta_i = k_i$ when $k_i \ll d$, and
- $\Delta_i = d - k_i$ when $d - k_i \ll d$.

We want to find all the tuples $(k_1, \ldots, k_n)$ for which the product of the terms $p_{k_i i}(x_i)$ corresponding to individual variables is larger than or equal to ε. The logarithm of the product is equal to the sum of the logarithms, so the logarithm if the product is asymptotically equal to $-\sum_{i=1}^{n} \Delta_i \cdot \ln(d)$. Thus, the condition that the product is larger than or equal to ε is asymptotically equivalent to the inequality

$$-\sum_{i=1}^{n} \Delta \cdot \ln(d) \geq \ln(\varepsilon),$$

i.e., to the inequality

$$\sum_{i=1}^{n} \Delta_i \leq C \overset{\text{def}}{=} \frac{|\ln(\varepsilon)|}{\ln(d)}.$$

The number of tuples of non-negative integers Δ_i that satisfy the inequality $\sum_{i=1}^{n} \Delta_i \leq C$ can be easily found from combinatorics.

Namely, we can describe each such tuple if we start with C zeros and then place ones:

- we place the first one after Δ_1 zeros,
- we place the second one after Δ_2 zeros following the first one,
- etc.

As a result, we get a sequence of $C + n$ symbols of which C are zeros. Vice versa, if we have a sequence of $C + n$ symbols of which C are zeros (and thus, n are ones), we can take:

- as Δ_1 the number of 0s before the first one,
- as Δ_2 the number of 0s between the first and the second ones,
- etc.

Thus, the total number of such tuples is equal to the number of ways that we can place C zeros in a sequence of $C + n$ symbols, i.e., equal to

$$\binom{C+n}{C} = \frac{(n+C) \cdot (n+C-1) \cdot \ldots \cdot (n+1)}{1 \cdot 2 \cdot \ldots \cdot C}.$$

When n is large, this number is asymptotically equal to

$$\text{const} \cdot n^C.$$

Each value Δ_i corresponds to two different values k_i:

- the value $k_i = \Delta_i$ and
- the value $k_i = d - k_i$.

Thus, to each tuple $(\Delta_1, \ldots, \Delta_n)$, there correspond 2^n different tuples $(k_1, \ldots, k_n)$. So, the total number of retained tuples $(k_1, \ldots, k_n)$ – i.e., tuples for which the largest value of the corresponding term is $\leq \varepsilon$ – is asymptotically equal to $2^n \cdot n^C$.

Conclusion: Bernstein Polynomials Are More Efficient Than Monomials. As we have shown:

- In the traditional computer representation of a polynomial of degree $\leq d$ in each of the variables, we need asymptotically $(d + 1)^n$ terms.
- For Bernstein polynomials, we need $2^n \cdot n^C$ terms.

For large n, the factor n^C grows much slower than the exponential term 2^n, and $2^n \ll (d + 1)^n$. Thus, in the Bernstein representation of a polynomial, we indeed need much fewer terms than in the traditional computer representation – and therefore, Bernstein polynomials are indeed more efficient.

3.6 How to Estimate the Accuracy of the Spatial Location of the Corresponding Measurement Results

Need for Gauging the Accuracy of Spatial Location. In practice, it is often important not only to describe the accuracy of the measurement result, but also to describe the accuracy with which we spatially locate these measurement results. We illustrate this problem on the example of seismic data processing. Before we explain the technical details, let us briefly describe this problem and explain why it is important.

Comment. This explanation first appeared in [76].

In Evaluations of Natural Resources and in the Search for Natural Resources, It Is Very Important to Determine Earth Structure. Our civilization greatly depends on the things we extract from the Earth, such as fossil fuels (oil, coal, natural gas), minerals, and water. Our need for these commodities is constantly growing, and because of this growth, they are being exhausted. Even under the best conservation policies, there is (and there will be) a constant need to find new sources of minerals, fuels, and water.

The only sure-proof way to guarantee that there are resources such as minerals at a certain location is to actually drill a borehole and analyze the materials extracted. However, exploration for natural resources using indirect means began in earnest during the first half of the 20th century. The result was the discovery of many large relatively easy to locate resources such as the oil in the Middle East.

However, nowadays, most easy-to-access mineral resources have already been discovered. For example, new oil fields are mainly discovered either at large depths, or under water, or in very remote areas – in short, in the areas where drilling is very expensive. It is therefore desirable to predict the presence of resources as accurately as possible before we invest in drilling.

From previous exploration experiences, we usually have a good idea of what type of structures are symptomatic for a particular region. For example, oil and gas tend to concentrate near the top of natural underground domal structures. So, to be able to distinguish between more promising and less promising locations, it is desirable to determine the structure of the Earth at these locations. To be more precise, we want to know the structure at different depths z at different locations (x, y).

Determination of Earth Structure Is Also Very Important for Assessing Earthquake Risk. Another vitally important application where the knowledge of the Earth structure is crucial is the assessment of earth hazards. Earthquakes can be very destructive, so it is important to be able to estimate the probability of an earthquake, where one is most likely to occur, and what will be the magnitude of the expected earthquake. Geophysicists have shown that earthquakes result from accumulation of mechanical stress; so if we know the detailed structure of the corresponding Earth locations, we can get a good idea of the corresponding stresses and faults present and the potential for occurrence of an earthquake. From this viewpoint, it is also very important to determine the structure of the Earth.

Data That We Can Use to Determine the Earth Structure. In general, to determine the Earth structure, we can use different measurement results that can be obtained without actually drilling the boreholes: e.g., gravity and magnetic measurements, analyzing the travel-times and paths of seismic ways as they propagate through the earth, etc.

Seismic Measurements Are Usually the Most Informative. Because of the importance and difficulty of the inverse problem, geophysicists would like to use all possible measurement results: gravity, magnetic, seismic data, etc. In this section, we concentrate on the measurements which carry the largest amount of information about the Earth structure and are, therefore, most important for solving inverse problems.

Some measurements – like gravity and magnetic measurements – describe the overall effect of a large area. These measurements can help us determine the average mass density in the area, or the average concentration of magnetic materials in the area, but they often do not determine the detailed structure of this area. This detailed structure can be determined only from measurements which are narrowly focused on small sub-areas of interest.

The most important of these measurements are usually *seismic measurements*. Seismic measurements involve the recording of vibrations caused by distant earthquakes, explosions, or mechanical devices. For example, these records are what seismographic stations all over the world still use to detect earthquakes. However, the signal coming from an earthquake carries not only information about the earthquake itself, it also carries the information about the materials along the path from an earthquake to the station: e.g., by measuring the travel-time of a seismic wave, checking how fast the signal came, we can determine the velocity of sound v in these materials. Usually, the velocity of sound increases with increasing density, so, by knowing the velocity of sound at different 3-D points, we can determine the density of materials at different locations and different depths.

The main problem with the analysis of earthquake data (i.e., *passive* seismic data) is that earthquakes are rare events, and they mainly occur in a few seismically active belts. Thus, we have a very uneven distribution of sources and receivers that results in a "fuzzy" image of earth structure in many areas.

To get a better understanding of the Earth structure, we must therefore rely on *active* seismic data – in other words, we must make artificial explosions, place sensors around them, and measure how the resulting seismic waves propagate. The most important information about the seismic wave is the *travel-time* t_i, i.e., the time that it takes for the wave to travel from its source to the sensor. to determine the geophysical structure of a region, we measure seismic travel times and reconstruct velocities at different depths from these data. The problem of reconstructing this structure is called the *seismic inverse problem*; see, e.g., [127].

How Seismic Inverse Problem Is Solved. First, we discretize the problem: we divide our 3-D spatial region into cells, and we consider the velocity values to be constant within each cell. The value of the velocity in the cell j is denoted by v_j.

Once we know the velocities v_j in each cell j, we can then determine the paths which seismic waves take. Seismic waves travel along the shortest path – shortest in terms of time. As a result, within each cell, the path is a straight line, and on the border between the two cells with velocities v and v', the direction of the path changes in accordance with Snell's law

$$\frac{\sin(\varphi)}{v} = \frac{\sin(\varphi')}{v'},$$

where φ and φ' are the angles between the paths and the line orthogonal to the border between the cells. (If this formula requires $\sin(\varphi') > 1$, this means that this wave cannot penetrate into the neighboring cell at all; instead, it bounces back into the original cell with the same angle φ.)

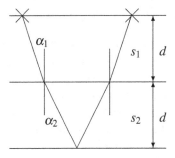

In particular, we can thus determine the paths from the source to each sensor. The travel-time t_i along i-th path can then be determined as the sum of travel-times in different cells j through which this path passes: $t_i = \sum_j \dfrac{\ell_{ij}}{v_j}$, where ℓ_{ij} denotes the length of the part of i-th path within cell j.

This formula becomes linear if we replace the original unknowns – velocities v_j – by their inverses $s_j \overset{\text{def}}{=} \dfrac{1}{v_j}$, called *slownesses*. In terms of slownesses, the formula for the travel-time takes the simpler form $t_i = \sum_j \ell_{ij} \cdot s_j$.

The system is not exactly linear, because the values ℓ_{ij} depend on the path and thus, depend on the velocities. To solve this problem, several methods have been proposed. One of the most popular methods, proposed by J. Hole in [60], consists of the following, We start with some initial reasonable values of velocities. Then, we repeat the following two steps until the process converges:

- based on the current values of the slownesses, we find the shortest pathes between sources and sensors and thus, the values ℓ_{ij};
- based on the current values ℓ_{ij}, we solve the above system of linear equations, and get the updated values of slownesses, etc.

Need to Find Spatial Resolution. Based on the seismic data, we produce a 3-D map describing the velocity v_j at different locations on different depths. Due to incomplete coverage and measurement uncertainty, this map provides only an approximate description of the actual velocity distribution. For example, based on the seismic data, it is impossible to distinguish between the densities at two nearby points. In other words, what we actually reconstruct is not a function of 3 variables, but rather values determined on the appropriate spatial granules. Because of this granularity, it is necessary to find the spatial resolution at different locations and at different depths, i.e., in other words, it is necessary to determine the corresponding granules.

Uncertainty vs. Spatial Resolution (Granularity). Actually, when we reconstruct the velocities in different cells, we have two types of uncertainty (see, e.g., [77, 132]:

- first, the "traditional" uncertainty – the reconstructed value of velocity is, due to measurement inaccuracy and incomplete coverage, only approximately equal to the actual (unknown) velocity value;
- second, the spatial resolution – each measured value represents not the value at a single point, but rather the "average" value over the whole region (*granule*) that affected the measurement; see, e.g., [128].

Methods of determining traditional uncertainty have been traditionally more developed, and the main ideas of methods for determined spatial resolution comes from these more traditional methods. In view of this, before we describe the existing methods for determining spatial resolution, let us describe the corresponding methods for determining more traditional uncertainty.

How Traditional Uncertainty Is Determined: Main Idea. There exist many techniques for estimating the "traditional" uncertainty; see, e.g., [3, 4, 28, 92] and references therein.

Most of these methods are based on the following idea. Usually, we know the accuracy of different measurements. We therefore

- first, we add, to the measured values of traveltimes, simulated noise of the size of the corresponding measurement errors; these values, due to our selection of noise, could possibly emerge if we simply repeat the same measurements;
- then, we reconstruct the new values of the velocities based on these modified traveltimes; these values come from realistic traveltimes and thus, can occur if we simply repeat the same measurements;
- finally, we compare the resulting velocities with the originally reconstructed ones: the difference between these two reconstructions is a good indication how accurate are these values.

Comment. Since the geophysical models involve a large amount of expert knowledge, it is also necessary to take into account the uncertainty of the expert statements; this is done, e.g., in [4, 5, 27].

How This Idea Is Applied to Determine Spatial Resolution. To determine spatial resolution, we can also simulate noise, the only difference is that this noise should reflect spatial resolution (granularity) and not the inaccuracy of the measurement values. Thus, we arrive at the following method:

- first, we add a perturbation of spatial size δ_0 (e.g., sinusoidal) to the reconstructed field $\widetilde{v}(x)$;
- then, we simulate the new traveltimes based on the perturbed values of the velocities;
- finally, we apply the same seismic data processing algorithm to the simulated traveltimes, and reconstruct the new field $\widetilde{v}_{\text{new}}(x)$.

If the perturbations are not visible in $\widetilde{v}_{\text{new}}(x) - \widetilde{v}(x)$, this means that details of spatial size δ_0 cannot be reconstructed. If perturbations are

- visible in one area of the map and
- not very clear in the other part of the map,

this means that

- in the first area, we can detect details with spatial resolution δ_0 while
- in the second area, the spatial resolution is much lower, and the details of this size are not visible.

In the geosciences, this method is known as a *checkerboard* method since adding 2-D sinusoidal periodic perturbations makes the map look like a checkerboard.

The use of this method to determine spatial resolution of seismic data processing is described, in detail, in [3, 77, 132, 135]. In particular, in [77, 132, 135], it is proven that the empirically optimal sinusoidal perturbations are actually optimal (under a reasonable formalization of the corresponding optimization problem).

Checkerboard Method: Main Limitation. The main limitation of the checkerboard method is that its running time is several times higher than the time of the original seismic data processing. Indeed,

- in addition to applying the seismic data processing algorithm to the original data,
- we also need to apply the same algorithm to the simulated data – and apply it several times, to make sure that we have reliable results about spatial resolution.

Seismic data processing is usually very time-consuming, often requiring hours and even days of computations on high performance computers. Thus, if we want to compute not only the 3-D maps themselves, but also the spatial resolution of the corresponding maps, the computation time drastically increases – and the whole process slows down.

It is therefore desirable to develop faster techniques for estimating spatial resolution of the corresponding maps, techniques that do not require new processing of simulated seismic data – and only use the results of the processing the original seismic data.

A Similar Problem Arises for Estimating Traditional Uncertainty. As we have mentioned, the existing methods for determining traditional uncertainty are also based on simulating errors and applying the (time-consuming) seismic data processing algorithms to the simulated traveltimes. As a result, the existing methods for determining the traditional uncertainty are also too time-consuming, and there is a similar need to developing faster uncertainty estimation techniques.

Since, as we mentioned, spatial resolution techniques usually emulate techniques for determining traditional uncertainty, let us therefore start with describing the existing techniques for faster

First Heuristic Idea for Estimating Uncertainty: Ray Coverage. In general, each measurement adds information about the affected quantities. The more measurements we perform, the more information we have and thus, the more accurately we can determine the desired quantity.

In particular, for each cell j, the value v_j affects those traveltime measurements t_i for which the corresponding path goes through this cell, i.e., for which $\ell_{ij} > 0$. Thus, the more rays pass through the cell, the more accurate the corresponding measurement. The number of such rays – called a *ray coverage* – is indeed reasonably well correlated with uncertainty and can, thus, serve as an estimate for this uncertainty:

- the smaller the ray coverage,
- the larger the uncertainty.

Limitations of Ray Coverage and the DWS Idea. Simply counting the ray does not take into account that some ray barely tough the cell, withe the values ℓ_{ij} very small. Clearly, such rays do not add much to the accuracy of determining the velocity v_j in the corresponding cell. It is therefore necessary to take into account not only how many rays go through the cell, but also how long are the paths of each ray in each cell. This idea was originally proposed by C. H. Thurber (personal communication, 1986) under the name of the Derivative Weight Sum; it was first published in [149].

As the name implies, instead of simply counting the rays that pass through a given cell j, we instead compute the sum of the lengths $D(j) = \sum_i \ell_{ij}$. This method have been successfully used in several geophysical problems; see, e.g., [140, 150, 152,

157]. It is indeed better correlated with the actual (simulation-computed) accuracy than the ray coverage.

Comment. The above form of the DWS is based on Hole's code approach, in which the slowness is assumed to be constant within each cell. An alternative approach is assuming that the slowness function is not piece-wise constant but rather piece-wise linear. In other words, we determine the values s_j at different points $j = (j_1, j_2, j_3)$ on a rectangular grid (here j_i are assumed to be integers), and we use linear extrapolation to describe the values $s(x_1, x_2, x_3)$ at arbitrary points $x_i = j_i + \alpha_i$ with $0 \le \alpha_i \le 1$.

In the 1-D case, linear interpolation takes the simple form $s(x) = \alpha \cdot s_{j+1} + (1 - \alpha) \cdot s_j$. To get the formula for the 2-D case, we first use linear interpolation to get the values

$$s(x_1, j_2) = \alpha_1 \cdot s_{j_1+1,j_2} + (1 - \alpha_1) \cdot s_{j_1,j_2}$$

and

$$s(x_1, j_2 + 1) = \alpha_1 \cdot s_{j_1+1,j_2+1} + (1 - \alpha_1) \cdot s_{j_1,j_2+1}$$

and then use 1-D linear interpolation to estimate $s(x_1, x_2)$ as

$$s(x_1, x_2) = \alpha_2 \cdot s(x_1, j_2 + 1) + (1 - \alpha_2) \cdot s(x_1, j_2),$$

i.e., substituting the above expressions for $s(x_1, j_2 + 1)$ and $s(x_1, j_2)$, the expression

$$s(x_1, x_2) = \alpha_1 \cdot \alpha_2 \cdot s_{j_1+1,j_2+1} + (1 - \alpha_1) \cdot \alpha_2 \cdot s_{j_1,j_2+1} +$$

$$\alpha_1 \cdot (1 - \alpha_2) \cdot s_{j_1+1,j_2} + (1 - \alpha_1) \cdot (1 - \alpha_2) \cdot s_{j_1,j_2}.$$

Similarly, we can go from the 2-D to the 3-D case, resulting in

$$s(x_1, x_2, x_3) = \alpha_3 \cdot s(x_1, x_2, j_3 + 1) + (1 - \alpha_3) \cdot s(x_1, x_2, j_3),$$

and

$$s(x_1, x_2, x_3) = \alpha_1 \cdot \alpha_2 \cdot \alpha_3 \cdot s_{j_1+1,j_2+1,j_3+1} + (1 - \alpha_1) \cdot \alpha_2 \cdot \alpha_3 \cdot s_{j_1,j_2+1,j_3+1} +$$

$$\alpha_1 \cdot (1 - \alpha_2) \cdot \alpha_3 \cdot s_{j_1+1,j_2,j_3+1} + (1 - \alpha_1) \cdot (1 - \alpha_2) \cdot \alpha_3 \cdot s_{j_1,j_2,j_3+1} +$$

$$\alpha_1 \cdot \alpha_2 \cdot (1 - \alpha_3) \cdot s_{j_1+1,j_2+1,j_3} + (1 - \alpha_1) \cdot \alpha_2 \cdot (1 - \alpha_3) \cdot s_{j_1,j_2+1,j_3} +$$

$$\alpha_1 \cdot (1 - \alpha_2) \cdot (1 - \alpha_3) \cdot s_{j_1+1,j_2,j_3} + (1 - \alpha_1) \cdot (1 - \alpha_2) \cdot (1 - \alpha_3) \cdot s_{j_1,j_2,j_3}.$$

Under this linear interpolation, in the formula for t_i, the coefficient at each term s_j is no longer ℓ_{ij}, but rather the integral of the corresponding interpolation coefficient $\omega_{ij}(x)$ over the ray path γ_i: $\int \omega_{ij}(x) d\gamma_i$. Thus, instead of the sum of the lengths, it is reasonable to take the sum of these integrals $D(j) = \sum_i \int \omega_{ij}(x) d\gamma_i$. This is actually the original form of the DWS.

Angular Diversity: A Similar Approach to Spatial Resolution. If we have many rays passing through the cell j, then we can find the slowness s_j in this cell with a high accuracy. However, if all these rays are parallel and close to each other, then all of them provide the information not about this particular *cell*, but rather about a *block of cells* following the common path. Thus, in effect, instead of the value s_j, we get the *average* value of slowness over several cells – i.e., we get a map with a low spatial resolution. To get a good spatial resolution, we must have "angular diversity", rays at different angles passing through the cell j.

The measure of such diversity called *ray density tensor* was proposed in [66]; see also [65, 141, 160]. In this measure, for each cell j, we form a 3×3 tensor (= matrix)

$$R_{ab}(j) = \sum_i \ell_{ij} \cdot e_{ij,a} \cdot e_{ij,b},$$

where $e_{ij} = (e_{ij,1}, e_{ij,2}, e_{ij,3})$ is unit vector in the direction in which the i-th ray crosses the j-th cell.

By plotting, for each unit vector $e = (e_1, e_2, e_3)$, the value $\sum_{a,b} R_{ab}(j) \cdot e_a \cdot e_b$ in the corresponding direction, we get an ellipsoid that describes spatial resolution in different directions. If this ellipsoid is close to a sphere, this means that we have equally good spatial resolution in different directions. If the ellipsoid is strongly tilted in one direction, this means that most of the ray are oriented in this direction, so spatial resolution in this direction is not good.

Limitations of the Known Approaches. From the application viewpoint, the main limitation is that these methods are, in effect, *qualitative*, in the following sense:

- the ray coverage, DWS, and the ray density tensor provide us with reasonable indications of where the uncertainty and/or spatial resolution are better and where they are worse;
- however, they do not give a geophysicist any specific guidance on how to use these techniques: what exactly is the accuracy? what exactly is the spatial resolution in different directions?

An additional limitation is that the above methods for gauging uncertainty and spatial resolution are *heuristic* techniques, they are not justified – statistically or otherwise.

It is therefore desirable to provide *justified quantitative* estimates for uncertainty and for spatial resolution.

Gauging Uncertainty: Gaussian Approach. For each cell j, each ray i that passes though it leads to an equation $\ell_{ij} \cdot s_j + \ldots = t_i$. If σ is the accuracy with which we measure each traveltime, then, in the assumption that the measurement errors are independent and normally distributed, the probability of a given value s_j is proportional to

$$\prod_i \exp\left(-\frac{(\ell_{ij} \cdot s_j + \ldots - t_i)^2}{2\sigma^2}\right).$$

By using the fact that $\exp(a) \cdot \exp(b) = \exp(a+b)$, we can represent this expression as $\sim \exp\left(-\frac{(s_j - \ldots)^2}{2\sigma(j)^2}\right)$, where $(\sigma(j))^2 \stackrel{\text{def}}{=} \frac{\sigma^2}{D_2(j)}$, and $D_2(j) \stackrel{\text{def}}{=} \sum_i \ell_{ij}^2$. Thus, the resulting estimate for s_j is normally distributed, with standard deviation $\sigma(j) = \frac{\sigma}{\sqrt{D_2(j)}}$.

This formula is similar for the formula for the DWS, with the only difference that we add up *squares* of the lengths instead of the lengths themselves.

Comment: fuzzy approach. We get the same expression if we use a fuzzy approach, with Gaussian membership functions and algebraic product $d \& d' = d \cdot d'$ as a t-norm.

In practice, we should consider *type-2* approaches, i.e., take into account that, e.g., the value σ is itself only known with uncertainty – e.g., we know only the interval $[\underline{\sigma}, \overline{\sigma}]$ of possible values of σ. In this case, we get $\underline{\sigma(j)} = \dfrac{\underline{\sigma}}{\sqrt{D_2(j)}}$ and

$$\overline{\sigma(j)} = \frac{\overline{\sigma}}{\sqrt{D_2(j)}}$$

Gauging Uncertainty: Robust Statistical Approach. In the case of the normal distribution, finding the most probable values of the parameters is equivalent to the Least Squares method $\sum_i \dfrac{e_i^2}{\sigma_i^2} \to \min$, where e_i is the difference between the model and measured values corresponding to the i-th measurement. Often, measurement and estimation errors are not normally distributed – and, moreover, we often do not know the shape of the corresponding distribution. In this case, instead of the Least Squares method corresponding to the normal distributions, it makes sense to consider so-called l^p-methods $\sum_i \dfrac{|e_i|^p}{\sigma_i^p} \to \min$ where a parameter p needs to be empirically determined; see, e.g., [28]. For seismic data processing, the empirical value p is close to 1; see [28].

The use of an l^p-method is equivalent to using the probability distribution $\sim \exp\left(-\dfrac{|e_i|^p}{\sigma^p}\right)$. For the seismic case $p = 1$ and $e_i = \ell_{ij} \cdot s_j + \ldots - t_i$, we thus get a term proportional to

$$\prod_i \exp\left(-\frac{|\ell_{ij} \cdot s_j + \ldots - t_i|}{\sigma}\right).$$

By combining the coefficients at s_j, we thus conclude that the standard deviation is approximately equal to $\sigma(j) = \dfrac{\sigma}{D(j)}$, where $D(j) = \sum_i \ell_{ij}$ is exactly the DWS expression. Thus, in the robust case, we get a statistical justification of the DWS formulas.

Comment. Similar formulas appear if, instead of Gaussian, we use exponential membership functions $\sim \exp(-c \cdot |e|)$. The uncertainty in σ can be handled similarly to the Gaussian case.

Gauging Spatial Resolution. What is the accuracy with which we can determine, e.g., the partial derivative $\dfrac{\partial s}{\partial x_1}$, i.e., in discrete terms, the difference $s' - s$, where $s' \overset{\text{def}}{=} s_{j_1+1, j_2, j_3}$ and $s \overset{\text{def}}{=} s_{j_1, j_2, j_3}$? This question is best answered in the above-described linear approximation. If the ray i is parallel to x_1 (i.e., $\alpha_{i1} = 0$), then, in the formula for t_i, the values s and s' are included with the same coefficient, so we can only determine the average value $(s+s')/2$.

In general, the difference between the corresponding interpolation coefficients at s and s' is proportional to $\ell_{ij} \cdot \sin(\alpha_{ij,1})$:

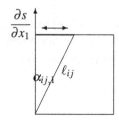

So, in addition to the term proportional to $(s+s')/2$, we also get a term proportional to $s' - s$, with a coefficient $\ell_{ij} \cdot \sin(\alpha_{ij,1})$. Similarly to the Gaussian approach to uncertainty, we can now argue that the accuracy with which we can determine the desired gradient is proportional to $\sigma_1^2 = \dfrac{\sigma^2}{D_{11}(j)}$, where $D_{11} = \sum\limits_i \ell_{ij}^2 \cdot \sin^2(\alpha_{ij,1})$. In vector terms, $\cos(\alpha_{ij,1}) = e_{ij,1}$, so $\sin^2(\alpha_{ij,1}) = 1 - e_{ij,1}^2$.

Thus, in general, the accuracy in the direction $e = (e_1, e_2, e_3)$ is $\sim \dfrac{\sigma}{\sqrt{D_e(j)}}$, where $D_e(j) = \sum D_{ab}(j) \cdot e_a \cdot e_b$, and

$$D_{ab}(j) = D_2(j) \cdot \delta_{ab} - \sum_i \ell_{ij}^2 \cdot e_{ij,a} \cdot e_{ij,b}.$$

This formula is similar to the ray density tensor formula, with ℓ_{ij}^2 instead of ℓ_{ij}. (In the robust case, we get exactly the ray density tensor.)

3.7 Is Feasible Prediction Always Possible?

Let Us Reformulate This Question in Precise Terms. In situation where prediction is possible, is it always possible to find a feasible algorithm for predicting future events? To make such a prediction, we need to know the corresponding physical laws. These laws can be found by analyzing the known observations and measurement results. Once the laws are found, predictions are usually reasonably easy, the big problem is finding the corresponding laws based on the known observations.

Once we have conjectured a possible set of physical laws, it is reasonably straightforward to check whether the existing observations satisfy these laws. For example, if we have conjectured that the current I, voltage V, and resistance R satisfy Ohm's law $V = I \cdot R$, we can simply look through all known observations in which we know the values of all these three quantities, and check whether the conjectured law holds for all these observations. The computation time for such a checking grows linearly with the size of the observations database. In other cases, we may need quadratic time – e.g., if we check that some property holds for every pair of observations – but in all these cases, the computation time needed for this checking is bounded by a polynomial of the size of the input data. In other words, if we made a guess, then checking whether this guess is correct can be done feasibly – in polynomial time.

In theoretical computer science, computations with guesses are called *non-deterministic*. In these terms, the problem of finding a physical law is non-deterministic polynomial, i.e., belongs to the class NP of all problems which can be solved in non-deterministic (= with guess) polynomial time. The question is whether we can have a feasible (polynomial-time) algorithm for solving all these problems, i.e., whether each of the problems belongs to the class P of all the problems that can solved feasibly (= in polynomial time). In other words, the question is whether every problem from the class NP also belongs to the class P, i.e., whether P=NP.

$P\overset{?}{=}$NP Problem: One of the Major Challenges in Theoretical Computer Science. The question of whether P=NP is one of longstanding open problems. In this section, we show that symmetries and similarities can provide a better understanding of this problem and thus, hopefully, contribute to the solution.

Comment. The main result of this session first appeared in [107].

Need for a Better Intuitive Understanding of the P=NP Option. In history of mathematics, solutions to many long-standing problems came when the consequences of the corresponding statements being true or false became clearer. For example, mathematicians have tried, for many centuries, to deduce the V-th Postulate – that for every point P outside a line ℓ, there is no more than one line ℓ' going through P and parallel to ℓ – from other postulates of geometry. The independence proof appeared only after the results of Gauss, Bolyai, and Lobachevsky made geometry without this postulate more intuitively clear; see, e.g., [16].

For this viewpoint, maybe one of the difficulties in solving the $P\overset{?}{=}$NP problem is that while there are good intuitive arguments in favor of P$\neq$NP, there is a definite lack of intuitively convincing arguments in favor of P=NP.

Example of Intuitive Arguments in Favor of P$\neq$NP. Example of arguments in favor of P$\neq$NP are numerous, many of them boil down to the following: if P=NP, it is possible to have feasible algorithms for solving classes of problems which are now considered highly creative – and for which, therefore, such algorithms are intuitively unlikely.

One example of a highly creative activity area is mathematics, where one of main objectives is, given a statement S, to prove either this statement or its negation $\neg$. We are usually interested in proofs which can be checked by human researchers, and are, thus, of reasonable size. In the usual formal systems of mathematics, the correctness of a formal proof can be checked in polynomial time. So, the problem of finding a reasonable-size proof of a given statement S (or of its negation) belongs to the class NP. If P was equal to NP, then we would be able to have a polynomial-time algorithm for solving all open problems of mathematics – a conclusion which most mathematicians consider unlikely.

Similarly, in theoretical physics, one of the main challenges is to find formulas that describe the observed data. The size of such a formula cannot exceed the amount of data, so the size is feasible. Once a formula is proposed, checking whether all the data is consistent with this formula is easy; thus, the problem of searching for such a formula is in the class NP. So, if P was equal to NP, we would have a feasible algorithm for the activity which is now considered one of the most creative ones – judged, e.g., by the fact that Nobel Prizes in Physics get a lot of publicity and bring a lot of prestige.

Physical Motivations: The Idea of Scale Invariance (Reminder). As we have mentioned in Chapter 1, the value of a physical quantity can be measured by using different units. For example, length can be measured in meters, in centimeters, in inches, etc. When we replace the original unit by a new unit which is λ times larger, all numerical values x change, from x to $x' = \dfrac{x}{\lambda}$, so that $x = \lambda \cdot x'$; this transformation is known as *re-scaling*.

For many physical processes, there is no preferred value of a physical quantity; see, e.g., [38]. For such processes, it is reasonable to require that the corresponding dependence have the same form no matter what measuring unit we use. For example, the dependence of the pendulum's period T on its length L has the form $T = f(L) = 2\pi \cdot \sqrt{\dfrac{L}{g}} = c \cdot \sqrt{L}$ for an appropriate constant c. If we change the unit of length, so that $L = \lambda \cdot L'$, we get a *similar* dependence $T = f(\lambda \cdot L') = c \cdot \sqrt{\lambda \cdot L'} = c \cdot \sqrt{\lambda} \cdot \sqrt{L'}$. If we now accordingly re-scale time, to new units which are $\sqrt{\lambda}$ times larger, then we get the exact *same* dependence in the new units $T' = c \cdot \sqrt{L'}$. Since we get the same formula for all measuring units, physicists say that the pendulum formula is *scale-invariant*.

In general, a dependence $y = f(x)$ is called scale-invariant if each re-scaling of x can be compensated by an appropriate re-scaling of y, i.e., if for every λ, there is a value $C(\lambda)$ for which $f(\lambda \cdot x) = C(\lambda) \cdot f(x)$ for all x and λ. For continuous functions, this functional equation leads to the power law $f(x) = c \cdot x^\alpha$; see, e.g., [1].

Scale-invariance is ubiquitous in physics: e.g., it helps explain most fundamental equations of physics, such as Einstein's equations of General Relativity, Schrödinger'e equations of quantum mechanics, Maxwell's equations, etc. [42]. It is also useful in explaining many semi-empirical computer-related formulas; see, e.g., [119].

Maybe Some Algorithms Are Scale-Invariant. One of the main concepts underlying P and NP is the concept of computational complexity $t_A(n)$ of an algorithm A, which is defined as the largest running time of this algorithm on all inputs of length $\leq n$. Similar to physics, in principle, we can use different units to measure the input length: we can use bits, bytes, Kilobytes, Megabytes, etc. It is therefore reasonable to conjecture that for some algorithms, the dependence $t_A(n)$ is scale-invariant – i.e., that its form does not change if we simply change a unit for measuring input length.

It should be mentioned that for discrete variables n, scale-invariance cannot be defined in exactly the same way as in physics, since the fractional length n/λ does not always make sense. Thus, we require scale-invariance only *asymptotically*, when $n \to \infty$:

Definition 3.7.1

- *We say that functions $f(n)$ and $g(n)$ are* asymptotically equivalent *(and denote it by $f(n) \sim g(n)$) if $f(n)/g(n) \to 1$ when $n \to \infty$.*
- *We say that a function $f(n)$ from natural numbers to natural numbers is* asymptotically scale-invariant *if for every integer k, there exists an integer $C(k)$ for which*

$$f(k \cdot n) \sim C(k) \cdot f(n).$$

- *We say that an algorithm A is* scale-invariant *if its computational complexity function $t_A(n)$ is scale-invariant.*

Now, we are ready to present the promised equivalent reformulation of P=NP, a reformulation that – in view of the ubiquity of scale invariance in physics – provides *some* intuitive argument in favor of this equality.

Proposition 3.7.1. P=NP *if and only if there exists a scale-invariant algorithm for solving propositional satisfiability SAT.*

Proof. If P=NP, then there exists a polynomial-time algorithm A for solving SAT, i.e., an algorithm for which $t_A(n) \leq C \cdot n^\alpha$ for some C and α. We can modify this algorithm as follows: first, we run A, then wait until the moment $C \cdot n^\alpha$. Thus modified algorithm A' also solves SAT, and its running time $t_{A'}(n) = C \cdot n^\alpha$ is clearly scale-invariant.

Vice versa, let us assume that A is a scale-invariant algorithm for solving SAT. For $k = 2$, this means that for some number $C(2)$, the ratio $\dfrac{t_A(2n)}{C(2) \cdot t_A(n)}$ tends to 1 as $n \to \infty$. By definition of the limit, that there exists an N such that for all $n \geq N$, we have $\dfrac{t_A(2n)}{C(2) \cdot t_A(n)} \leq 2$, i.e., $t_A(2n) \leq 2 \cdot C(2) \cdot t_A(n)$. By induction, for values $n = 2^k \cdot N$, we can now prove that $t_A(2^k \cdot N) \leq (2 \cdot C(2))^k \cdot t_A(N)$.

For every $n \geq N$, the smallest k for which $2^k \cdot N \geq n$ can be found as $k = \lceil \log_2(n/N) \rceil \leq \log_2(n/N) + 1$. By definition, the function $t_A(n)$ is non-decreasing, hence $t_A(n) \leq t_A(2^k \cdot N)$ and thus, $t_A(n) \leq (2 \cdot C(2))^k \cdot t_A(N)$. Due to the above inequality for k, we get

$$t_A(n) \leq (2 \cdot C(2))^{\log_2(n/N)+1} \cdot t_A(N) = (2 \cdot C(2))^{\log_2(n/N)} \cdot 2 \cdot C(2) \cdot f(N).$$

Here,

$$(2 \cdot C(2))^{\log_2(n/N)} = \left(2^{\log_2(2 \cdot C(2))}\right)^{\log_2(n/N)} = 2^{\log_2(2 \cdot C(2)) \cdot \log_2(n/N)} =$$

$$\left(2^{\log_2(n/N)}\right)^{\log_2(2 \cdot C(2))} = \left(\frac{n}{N}\right)^\alpha,$$

where $\alpha \stackrel{\text{def}}{=} \log_2(2 \cdot C(2))$, so $t_A(n) \leq \left(\dfrac{n}{N}\right)^\alpha \cdot 2 \cdot C(2) \cdot f(N)$. Thus, the SAT-solving algorithm A is indeed polynomial time, and hence, P=NP. The proposition is proven.

Chapter 4
Algorithmic Aspects of Control: Approach Based on Symmetry and Similarity

4.1 Algorithmic Aspects of Control

In the previous chapter, we concentrated on the problem of predicting the future events. Once we are able to predict future events, a natural next step is to *influence* these events, i.e., to *control* the corresponding system. In this step, we should select a control that leads to the best possible result.

Control problems can be roughly classified into two classes. In some problems of this type, we know the exact equations and we know the objective function that describes what the users want. In such problems, the selection of the best possible control is a mathematically well-defined optimization problem. Problems of this type have been solved for centuries, and there are many efficient algorithms that solve these types of problems in many practical situations.

However, there are situations in which the problems of finding the best control are much more challenging, because we only know the system with a huge uncertainty. Because of the uncertainty, it is difficult to formulate the corresponding problem as a precise optimization problem. Instead, we use intelligent techniques that use the knowledge and experience of human experts in solving such problems. Such intelligent techniques are reasonably new: they have been in use for only a few decades. Details of many of such techniques have been determine purely *empirically*, and as a result, they often lead to results which are far from optimal. To improve the results of applying these techniques, it is therefore imperative to perform a *theoretical* analysis of the corresponding problems. In this chapter, in good accordance with the general ideas from Chapter 1, we show that techniques based on similarity and symmetry can be very useful in this theoretical analysis.

We illustrate this usefulness on all level of the problem of selecting the best control. First, on a strategic level, we need to select the best class of strategies. In Section 4.2, we use logical symmetries – the symmetry between true and false values – to find the best class of strategies for an important class of intelligent controls – fuzzy control.

Once a class of strategies is selected, we need to select the best strategy within a given class. We analyze this problem in Sections 4.3 and 4.4: in Section 4.3, we use

© Springer-Verlag Berlin Heidelberg 2015
J. Nava and V. Kreinovich, *Algorithmic Aspects of Analysis, Prediction, and Control in Science and Engineering*, Studies in Systems, Decision and Control 14, DOI: 10.1007/978-3-662-44955-4_4

approximate symmetries (similarities) to find the best operations for implementing fuzzy control, and in Section 4.4, again in good accordance with Chapter 1, that the optimal selection of operations leads to a solution based on symmetry and similarity.

Finally, when we have several strategies coming from different aspects of the problem, we need to combine these strategies into a single strategy that takes all the aspects into account. In Section 4.5, we again use logical symmetries – this time to find the best way of combining the resulting fuzzy decisions.

Overall, we show that symmetries and similarities can help with all the algorithmic aspects of control.

4.2   Selecting the Best Class of Strategies: Case of Intelligent Control

Need for Fuzzy Control. As we have mentioned in the previous section, in many application areas, we do not have the exact control strategies, but we have human operators who are skilled in the corresponding control. Human operators are often unable to describe their knowledge in a precise quantitative form. Instead, they describe their knowledge in terms of control rules, rules that formulate their expertise by using words from natural language. These rules usually have the form "If $A_i(x)$ then $B_i(u)$", they relate properties of the input x with properties of the corresponding control u. For example, a rule may say "If a car in front is somewhat too close, break a little bit".

Fuzzy control is a set of techniques for transforming these rules into a precise control strategy; see, e.g., [70, 120].

Mamdani Approach to Fuzzy Control. Historically the first – and still most widely used – idea of fuzzy control was described by E. Mamdani. Mamdani argued that for a given input x, a control value u is reasonable if:

- either the first rule is applicable, i.e., its condition $A_1(x)$ is satisfied and its conclusion $B_1(u)$ is satisfied,
- or the second rule is applicable, i.e., its condition $A_2(x)$ is satisfied and its conclusion $B_2(u)$ is satisfied,
- etc.

Thus, in Mamdani's approach, the condition $R(x,u)$ meaning that the control u is reasonable for the input x takes the following form

$$(A_1(x)\,\&\,B_1(u)) \vee (A_2(x)\,\&\,B_2(u)) \vee \ldots \tag{4.1}$$

For a given input x_0, to get a desired control value $u(x_0)$, we must now apply an appropriate defuzzification procedure to the resulting membership function $R(x_0,u)$.

Logical Approach to Fuzzy Control. An alternative (more recent) approach to fuzzy control is to simply state that all the rules are valid, i.e., that the following statement holds:

$$(A_1(x) \to B_1(u)) \,\&\, (A_2(x) \to B_2(u)) \,\&\, \ldots \tag{4.2}$$

For example, we can interpret $A \to B$ as $\neg A \vee B$, in which case the formula (4.2) has the form

$$(\neg A_1(x) \vee B_1(u)) \,\&\, (\neg A_2(x) \vee B_2(u)) \,\&\, \ldots, \tag{4.3}$$

or, equivalently, the form

$$(A_1'(x) \vee B_1(u)) \,\&\, (A_2'(x) \vee B_2(u)) \,\&\, \ldots, \tag{4.4}$$

where $A_i'(x)$ denotes $\neg A_i(x)$.

Both Approaches Have a Universality Property. Both Mamdani's and logical approaches to fuzzy control have a universality (universal approximation) property [73, 120, 133] meaning that an arbitrary control strategy can be, with arbitrary accuracy, approximated by controls generated by this approach.

Corresponding Crisp Universality Property. One of the reasons why the corresponding fuzzy controls have the universal approximation property is that the corresponding crisp formulas (4.1) and (4.2) have the following universal property: for finite sets X and U, an arbitrary relation $C(x, u)$ on $X \times U$ can be represented both in the form (4.1) and in the form (4.2), for appropriate properties $A_i(x)$ and $B_i(u)$.

Indeed, an arbitrary crisp property $C(x, u)$ can be described by the set $C \subseteq X \times U$ of all the pairs (x, u) that satisfy this property. Once this set is given, we can represent the corresponding property in the form (4.1) by taking

$$C(x, u) \Leftrightarrow \vee_{(x_0, u_0) \in C}((x = x_0) \,\&\, (u = u_0)) \tag{4.5}$$

and in the form (4.2) (equivalent to (4.4)) by taking

$$C(x, u) \Leftrightarrow \&_{(x_0, u_0) \notin C}((x = x_0) \to (u \neq u_0)). \tag{4.6}$$

Comment. This universality property is well known and actively used, e.g., in digital design: when we design, e.g., a vending machine, then to implement a general logical condition in terms of "and", "or", and "not"-gates, we first represent this condition in Conjunctive Normal Form (CNF) or in a Disjunctive Normal Form (DNF). These forms correspond exactly to our formulas (4.1) and (4.4) (equivalent to (4.2)), and the possibility to transform each logical condition into one of these forms is our universality property.

Fuzzy Control: What Other Approaches Are Possible? Both Mamdani's and logical approaches are actively used in fuzzy control. The fact that both approaches are actively used means that both have advantages and disadvantages, i.e., none of these two approaches is perfect. Since both are not perfect, it is reasonable to analyze what other approaches are possible.

In this section, we start this analysis by analyzing what type of crisp forms like (4.1) and (4.2) are possible.

Comment. The main result of this section first appeared in [19].

Definitions and the Main Result. In the above two representations, we used &, ∨, and →. These logical connectives are examples of *binary operations* in the following precise sense.

Definition 4.2.1. *By a* binary operation, *we mean a function*

$$f : \{0,1\} \times \{0,1\} \to \{0,1\}$$

that transforms two Boolean values a and b into a new Boolean value f(a,b).

Comment. In this section, as usual, we identify "false" with 0 and "true" with 1. We are looking for general representations of the type

$$(A_1(x) \odot B_1(u)) \ominus (A_2(x) \odot B_2(u)) \ominus \ldots, \qquad (4.7)$$

for arbitrary pairs of binary operations; we denoted these general binary operations $\odot$ and $\ominus$. For example, in the above representations, we used $\ominus = \vee$ and $\ominus = \&$; we want to find all other binary operations for which such a representation is possible.

It is important to notice that the operation $\ominus$ is used to combine different rules. Therefore, the result of this operation should not depend on the order in which we present the rules. Thus, this operation should be commutative and associative.

So, we arrive at the following definitions.

Definition 4.2.2. *We say that a pair of binary operations $(\odot, \ominus)$ in which the operation $\ominus$ is commutative and associative has a* universality property *if for every two finite sets X and Y, an arbitrary relation C(x,u) can be represented in the form (4.7) for appropriate relations $A_i(x)$ and $B_i(u)$.*

Discussion. One can easily check that if the pair $(\odot, \ominus)$ has the universality property, then the pair $(\odot', \ominus)$, where $a \odot' b \overset{\text{def}}{=} \neg a \odot b$, also has the universality property: indeed, each statement of the type $A_i(x) \odot B_i(u)$ can be equivalently represented as $A'_i(x) \odot' B_i(u)$ for $A'_i(x) \overset{\text{def}}{=} \neg A_i(x)$.

Similarly, if the pair $(\odot, \ominus)$ has the universality property, then the pair $(\odot', \ominus)$, where $a \odot' b \overset{\text{def}}{=} a \odot \neg b$, also has the universality property: indeed, each statement of the type $A_i(x) \odot B_i(u)$ can be equivalently represented as

$$A_i(x) \odot' B'_i(u)$$

for $B'_i(u) \overset{\text{def}}{=} \neg B_i(u)$.

Finally, if the pair $(\odot, \ominus)$ has the universality property, then the pair $(\odot', \ominus)$, where $a \odot' b \overset{\text{def}}{=} \neg a \odot \neg b$, also has the universality property: indeed, each statement of the type $A_i(x) \odot B_i(u)$ can be equivalently represented as

$$A'_i(x) \odot' B'_i(u)$$

for $A'_i(x) \overset{\text{def}}{=} \neg A_i(x)$ and $B'_i(u) \overset{\text{def}}{=} \neg B_i(u)$.

Thus, from the viewpoint of universality, the relations $\odot$ and $\odot'$ are equivalent. So, we arrive at the following definition.

Definition 4.2.3. *We say that binary operations $\odot$ and $\odot'$ are* similar *if the relation $\odot'$ has one of the following forms:*

$$a \odot' b \overset{\text{def}}{=} \neg a \odot b, \ a \odot' b \overset{\text{def}}{=} a \odot \neg b, \ \text{or } a \odot' b \overset{\text{def}}{=} \neg a \odot \neg b.$$

Definition 4.2.4. *We say that pairs $(\odot, \ominus)$ and $(\odot', \ominus)$ are* similar *if the operations $\odot$ and $\odot'$ are equivalent.*

The above discussion can be formulated as follows:

Proposition 4.2.1. *If the binary operations $\odot$ and $\odot'$ are similar, then the following two statements are equivalent to each other:*

- *the pair $(\odot, \ominus)$ has the universality property;*
- *the pair $(\odot', \ominus)$ has the universality property.*

Comment. One can easily check that the similarity relation is symmetric and transitive, i.e., in mathematical terms, that it is an equivalent relation. Thus, to classify all pairs with the universality property, it is sufficient to consider equivalence classes of binary operations $\odot$ with respect to the similarity relation.

Discussion. Our definition of a universal property requires that the rule-combining operation be commutative and associative. It turns out that there are only three such operations.

Proposition 4.2.2. *Out of all binary operations, only the following six are commutative and associative:*

- *the "zero" operation for which $f(a,b) = 0$ for all a and b;*
- *the "one" operation for which $f(a,b) = 1$ for all a and b;*
- *the "and" operation for which $f(a,b) = a \& b$;*
- *the "or" operation for which $f(a,b) = a \vee b$;*
- *the "exclusive or" operation for which $f(a,b) = a \oplus b$;*
- *the operation $a \oplus' b \overset{\text{def}}{=} a \oplus \neg b$.*

Comments

- The proof of this proposition is given in the following section.
- The "exclusive or" operation is actively used in digital design: e.g., when we add two binary numbers which end with digits a and b, the last digit of the sum is $a \oplus b$ (and the carry is $a \& b$). In view of this, "exclusive or" is also called *addition modulo 2*.
- Due to Proposition 4.2.2, it is sufficient to consider only these six operations $\ominus$. The following theorem provides a full classification of all such operations.

Theorem 4.2.1. *Every pair of operations with the universality property is similar to one of the following pairs:* $(\vee, \&)$, $(\&, \vee)$, $(\oplus, \vee)$, $(\oplus, \&)$, $(\oplus', \vee)$, $(\oplus', \&)$, *and all these six pairs of operations have the universality property.*

Discussion. Thus, in addition to the Mamdani and logical approaches, we have four other possible pairs with the universality property.

What is the meaning of the four additional pairs of operations? As we can see from the proof, for each operation $\odot$, the combination

$$(A_1(x) \odot B_1(u)) \oplus' (A_2(x) \odot B_2(u)) \oplus' \ldots \oplus' (A_n(x) \odot B_n(u))$$

is equal to

$$(A_1(x) \odot B_1(u)) \oplus (A_2(x) \odot B_2(u)) \oplus \ldots (A_n(x) \odot B_n(u))$$

for odd n and to the negation of this relation for even n.

Thus, for odd n, the use of operation $\oplus'$ to combine the rules is equivalent to using the original "exclusive or" operation $\oplus$. For even n, in order to represent an arbitrary property $C(x,u)$, we can use the $\oplus$ combination to represent its negation $\neg C(x,u)$ – this is equivalent to representing the original property by $\oplus'$.

Thus, in essence, in addition to forms (4.1) and (4.2), we only have two more forms:

$$(A_1(x) \& B_1(u)) \oplus (A_2(x) \& B_2(u)) \oplus \ldots \qquad (4.8)$$

and

$$(A_1(x) \vee B_1(u)) \oplus (A_2(x) \vee B_2(u)) \oplus \ldots \qquad (4.9)$$

The meaning of these forms is that, crudely speaking, we restrict ourselves to the cases where exactly one rule is applicable. The case of *fuzzy transforms (f-transforms*, for short) [130, 131]), where we consider rules "if $A_i(x)$ then $B_i(u)$" for which $\sum_{i=1}^{n} A_i(x) = 1$, can be therefore viewed as a natural fuzzy analogue of these cases.

Proofs

$1°$. In order to prove Proposition 4.2.2 and Theorem 4.2.1, let us first recall all possible binary operations. By definition, to describe a binary operation, one needs to describe four Boolean values: $f(0,0)$, $f(0,1)$, $f(1,0)$, and $f(1,1)$. Each of these four quantities can have two different values: 0 and 1; thus, totally, we have $2^4 = 16$ possible operations.

A natural way to classify these operations is to describe how many 1s we have as values $f(a,b)$. Out of 4 values, we can have 0, 1, 2, 3, and 4 ones. Let us describe these cases one by one.

$1.1°$. When we have zero 1s, this means that all the values $f(a,b)$ are zeros. Thus, in this case, we have a binary operation that always returns zero: $f(a,b) = 0$ for all a and b. It is easy to show that this operation cannot lead to the universality property:

- if we use this operation as $\odot$, then the formula (4.7) turns into a constant $0 \ominus 0 \ominus$ ... independent on x and u; thus, it cannot have the universality property;
- if we use this operation as $\ominus$, then the formula (4.7) turns into a constant 0, and thus, also cannot have the universality property.

$1.2°$. Similarly, when we have four 1s, this means that $f(a,b) = 1$ for all a and b, and we do not have a universality property.

$1.3°$. When we have a single one, this means that we have an operation similar to "and". Indeed, if $f(1,1) = 1$ and all other values $f(a,b)$ are 0s, this means that $f(a,b)$ is true if and only if a is true and b is true, i.e., that $f(a,b) \Leftrightarrow a \& b$. Similarly, if $f(1,0) = 1$, then $f(a,b) \Leftrightarrow a \& \neg b$; if $f(0,1) = 1$, then $f(a,b) \Leftrightarrow \neg a \& b$; and if $f(0,0) = 1$, then $f(a,b) \Leftrightarrow \neg a \& \neg b$.

$1.4°$. Similarly, we can prove that when we have three ones, this means that we have an operation similar to "or".

$2°$. To complete our classification, it is sufficient to describe all the cases where we have exactly two 1s. By enumerating all possible binary operations, we can check that in this case, we have six options: $f(a,b) = a$, $f(a,b) = \neg a$, $f(a,b) = b$, $f(a,b) = \neg b$, $a \oplus b$, and $a \oplus \neg b$. By analyzing these operations $f(a,b)$ one by one and by testing commutativity $f(a,b) = f(b,a)$ and associativity $f(a,f(b,c)) = f(f(a,b),c)$ for all possible values a, b, and c, we can describe all commutative and associative operations – i.e., prove Proposition 4.2.2.

$3°$. Arguments similar to the ones that we just gave enables us to prove the statement listed after the formulation of Theorem 4.2.1: that for any statements $S_1, \ldots, S_n$:

- for odd n, we have

$$S_1 \oplus' \ldots \oplus' S_n \Leftrightarrow S_1 \oplus \ldots \oplus S_n;$$

- for even n, we have

$$S_1 \oplus' \ldots \oplus' S_n \Leftrightarrow \neg(S_1 \oplus \ldots \oplus S_n).$$

Indeed, as we have mentioned, $S \oplus' S'$ is equivalent to $S \oplus S' \oplus 1$. Thus, for every n, due to associativity and commutativity of both operations $\oplus$ and $\oplus'$, we have:

$$S_1 \oplus' S_2 \oplus' S_3 \oplus' \ldots \oplus' S_n \Leftrightarrow (\ldots((S_1 \oplus' S_2) \oplus' S_3) \oplus' \ldots \oplus' S_n \Leftrightarrow$$

$$(\ldots((S_1 \oplus S_2 \oplus 1) \oplus S_3 \oplus 1) \oplus \ldots \oplus S_n \oplus 1 \Leftrightarrow$$

$$S_1 \oplus S_2 \oplus 1 \oplus S_2 \oplus 1 \oplus \ldots \oplus S_n \oplus 1 \Leftrightarrow$$

$$(S_1 \oplus S_2 \oplus S_3 \oplus \ldots \oplus S_n) \oplus (1 \oplus 1 \oplus \ldots \oplus 1)(n-1 \text{ times}).$$

Here, $1 \oplus 1 = 0$, So, for odd n, when $n - 1$ is even, we have

$$(1 \oplus 1 \oplus \ldots \oplus 1)(n - 1 \text{ times}) = (1 \oplus 1) \oplus (1 \oplus 1) \oplus \ldots \oplus (1 \oplus 1) =$$

$$0 \oplus 0 \oplus \ldots \oplus 0 = 0$$

and thus

$$S_1 \oplus' S_2 \oplus' S_3 \oplus' \ldots \oplus' S_n \Leftrightarrow$$

$$(S_1 \oplus S_2 \oplus S_3 \oplus \ldots \oplus S_n) \oplus (1 \oplus 1 \oplus \ldots \oplus 1)(n-1 \text{ times}) \Leftrightarrow$$

$$S_1 \oplus S_2 \oplus S_3 \oplus \ldots \oplus S_n.$$

Similarly, for even n, when $n-1$ is odd, we have

$$(1 \oplus 1 \oplus \ldots \oplus 1)(n-1 \text{ times}) = (1 \oplus 1) \oplus (1 \oplus 1) \oplus \ldots \oplus (1 \oplus 1) \oplus 1 =$$

$$0 \oplus 0 \oplus \ldots \oplus 0 \oplus 1 = (0 \oplus 0 \oplus \ldots \oplus 0) \oplus 1 = 0 \oplus 1 = 1$$

and thus

$$S_1 \oplus' S_2 \oplus' S_3 \oplus' \ldots \oplus' S_n \Leftrightarrow$$

$$(S_1 \oplus S_2 \oplus S_3 \oplus \ldots \oplus S_n) \oplus (1 \oplus 1 \oplus \ldots \oplus 1)(n-1 \text{ times}) \Leftrightarrow$$

$$(S_1 \oplus S_2 \oplus S_3 \oplus \ldots \oplus S_n) \oplus 1 \Leftrightarrow \neg(S_1 \oplus S_2 \oplus S_3 \oplus \ldots \oplus S_n).$$

4°. Due to Proposition 4.2.2, we only have to consider the six operations $\ominus$ described in this Proposition when checking the universality property.

We have already shown, in Part 1 of this proof, that pairs with $\ominus = 0$ and $\ominus = 1$ do not have the universality property.

We have also shown that a pair $(\odot, \oplus')$ has a universality property if and only if the pair $(\odot, \oplus)$ has the universality property.

Thus, it is sufficient to consider only three possible operations $\ominus$: $\&$, $\vee$, and $\oplus$. In the following text, we analyze these three cases one by one, and for each of these three cases, we consider all possible operations $\odot$.

5°. For $\odot$, as we have mentioned in the discussion from the main text, it is sufficient to consider only one operation from each class of operations which are similar to each other. According to the general classification (Part 1 of this proof), this leaves us with the operations 0, 1, $\&$, $\vee$, $\oplus$, and the degenerate operations – i.e., operations $f(a,b)$ which only depend on one of the two variables a or b.

5.1°. We have already shown (in Part 1) that we cannot have $\odot = 0$ or $\odot = 1$. Thus, for $\odot$, we have to consider cases where $\odot = \&$, where $\odot = \vee$, where $\odot = \oplus$, and where $a \odot b$ is one of the four "degenerate" operations $a \odot b = a$, $a \odot b = \neg a$, $a \odot b = b$, and $a \odot b = \neg b$.

5.2°. Let us prove that the pairs $(\odot, \ominus)$ for which $a \odot b = a$, $a \odot b = \neg a$, $a \odot b = b$, or $a \odot b = \neg b$, cannot have the universality property.

Indeed, e.g., for $a \odot b = a$, each expression $A_i(x) \odot B_i(u)$ has the form $A_i(x)$. Thus, the $\ominus$-combination of these expressions $A_1(x) \ominus A_2(x) \ominus \ldots \ominus A_n(x)$ does not depend on u at all and thus, cannot represent any property $C(x,u)$ that actually depends on u. Similarly, for $a \odot b = \neg a$, we get the expression $\neg A_1(x) \ominus \neg A_2(x) \ominus \ldots \ominus \neg A_n(x)$ which also does not depend on u and thus, cannot represent any property $C(x,u)$ that actually depends on u.

For $a \odot b = b$ and $a \odot b = \neg b$, we get, correspondingly, expressions

$$B_1(u) \ominus B_2(u) \ominus \ldots \ominus B_n(u)$$

and $\neg B_1(u) \ominus \neg B_2(u) \ominus \ldots \ominus \neg B_n(u)$ which do not depend on x and thus, cannot represent any property $C(x, u)$ that actually depends on x.

5.3°. Because of what we have proved in Parts 5.1 and 5.2, it is sufficient to consider only three operations $\odot$ for combining the premise $A_i(x)$ and the conclusion $B_i(u)$ of each rule: &, $\vee$, and $\oplus$.

6°. Let us first consider the case where $\ominus = \&$. In accordance with Part 5 of this proof, it is sufficient to analyze the universality property for the three subcases where $\odot = \&$, $\odot = \vee$, and $\odot = \oplus$. Let us consider these subcases one by one.

6.1°. When $\ominus = \&$ and $\odot = \&$, the general expression

$$(A_1(x) \odot B_1(u)) \ominus (A_2(x) \odot B_2(u)) \ominus \ldots$$

takes the form

$$(A_1(x) \& B_1(u)) \& (A_2(x) \& B_2(u)) \& \ldots$$

Due to commutativity and associativity of the "and" operation, this expression is equivalent to

$$(A_1(x) \& A_2(x) \& \ldots) \& (B_1(u) \& B_2(u) \& \ldots),$$

i.e., to $A(x) \& B(u)$, where $A(x) \overset{\text{def}}{=} A_1(x) \& A_2(x) \& \ldots$ and

$$B(u) \overset{\text{def}}{=} B_1(u) \& B_2(u) \& \ldots.$$

One can easily see that not every property $C(x, u)$ can be represented as $A(x) \& B(u)$. Indeed, let us take arbitrary sets X and U with at least two elements each, and let $x_0 \in X$ and $u_0 \in U$ be arbitrary elements from these sets. Let us prove, by contradiction, that the property $(x = x_0) \vee (u = u_0)$ cannot be represented in the form $A(x) \& B(u)$. Indeed, let us assume that for some properties $A(x)$ and $B(u)$, for every $x \in X$ and $u \in U$, we have

$$((x = x_0) \vee (u = u_0)) \Leftrightarrow (A(x) \& B(u)). \tag{4.10}$$

In particular, for $x = x_0$ and $u = u_1 \neq u_0$, the left-hand side of this equivalence (4.10) is true, hence the right-hand side $A(x_0) \& B(u_1)$ is true as well. Thus, both statements $A(x_0)$ and $B(u_1)$ are true.

Similarly, for $x = x_1 \neq x_0$ and $u = u_0$, the left-hand side of the equivalence (4.10) is true, hence the right-hand side $A(x_1) \& B(u_0)$ is true as well. Thus, both statements $A(x_1)$ and $B(u_0)$ are true.

Since $A(x_1)$ and $B(u_1)$ are both true, the conjunction $A(x_1) \& B(u_1)$ is also true, so due to (4.10), we would conclude that $(x_1 = x_0) \vee (u_1 = u_0)$, which is false. The contradiction proves that the representation (4.10) is indeed impossible and thus, the pair $(\&, \&)$ does not have the universality property.

6.2°. For $\ominus = \&$ and $\odot = \vee$, the universality property is known to be true – this is one of the two basic cases with which we started our analysis.

6.3°. Let us prove that the subcase $\ominus = \&$ and $\odot = \oplus$ does not lead to the universality property.

In this case, the general expression

$$(A_1(x) \odot B_1(u)) \ominus (A_2(x) \odot B_2(u)) \ominus \ldots$$

takes the form

$$(A_1(x) \oplus B_1(u)) \& (A_2(x) \oplus B_2(u)) \& \ldots \tag{4.11}$$

Let us prove, by contradiction, that for every $x_0 \in X$ and $u_0 \in U$, the property $C(x, u) \Leftrightarrow x \neq x_0 \vee u \neq u_0$ cannot be represented in the form (4.11). Indeed, let us assume that this representation is possible, for some properties $A_i(x)$ and $B_i(u)$.

For the above property $C(x, u)$, the set S of all the values for which this property is true contains all the pairs $(x, u) \in X \times U$ except for the pair (x_0, u_0). Due to equivalence, this same set S is also the set of all the pairs for which the formula (4.11) holds.

Due to the known properties of the "and" operations, the set S of all the values (x, u) for which the formula (4.11) holds is equal to the intersection of the sets $S_i = \{(x, u) : A_i(x) \oplus B_i(u)\}$. Thus, each of the sets S_i is a superset of the set S: $S \subseteq S_i$. By our construction, the set S is missing only one element; thus, it has only two supersets: itself and the set $X \times U$ of all the pairs. If all the sets S_i coincided with $X \times U$, then their intersection would also be equal to $X \times U$, but it is equal to $S \neq X \times U$. Thus, at least for one i, we have $S_i = S$. For this i, we have the equivalence

$$((x \neq x_0) \vee (u \neq u_0)) \Leftrightarrow (A_i(x) \oplus B_i(u)). \tag{4.12}$$

Let us now reduce this equivalence to the case where $A_i(x_0)$ is true (i.e., where $A_i(x_0) = 1$). Specifically, if $A_i(x_0)$ is false ($A_i(x_0) = 0$), then, since $A \oplus B \Leftrightarrow \neg A \oplus \neg B$, we can replace the original equivalence with the new one

$$((x \neq x_0) \vee (u \neq u_0)) \Leftrightarrow (A_i'(x) \oplus B_i'(u)),$$

where $A_i'(x) \overset{\text{def}}{=} \neg A_i(x)$ and $B_i'(u) \overset{\text{def}}{=} \neg B_i(u)$, and $A_i'(x_0) = \neg A_i(x_0) =$"true". So, we can assume that $A_i(x_0) = 1$.

Now, for $x = x_0$ and $u = u_0$, the left-hand side of the equivalence (4.12) is false, hence the right-hand side $A_i(x_0) \oplus B_i(u_0)$ is false as well. Since we assumed that $A_i(x_0) = 1$, by the properties of "exclusive or", we thus conclude that $B_i(u_0) = 1$.

For $x = x_1 \neq x_0$ and $u = u_0$, the left-hand side of the equivalence (4.12) is true, hence the right-hand side $A_i(x_1) \oplus B_i(u_0)$ is true as well. Since, as we have already proven, $B_i(u_0)$ is true, we conclude that $A_i(x_1)$ is false.

Similarly, for $x = x_0$ and $u = u_1 \neq u_0$, the left-hand side of the equivalence (4.12) is true, hence the right-hand side $A_i(x_0) \oplus B_i(u_1)$ is true as well. Since $A_i(x_0)$ is true, we conclude that $B_i(u_1)$ is false.

Now, for $x = x_1$ and $u = u_1$, both formulas $A_i(x_1)$ and $B_i(u_1)$ are false, hence their combination $A_i(x_1) \oplus B_i(u_1)$ is also false. So, due to (4.12), we would conclude that the statement $(x_1 \neq x_0) \vee (u_1 \neq u_0)$ is false, but this statement is actually true. The contradiction proves that the representation (4.12) is indeed impossible, and so, the pair $(\oplus, \&)$ does not have the universality property.

$7°$. Let us now consider the case where $\ominus = \vee$. In accordance with Part 5 of this proof, it is sufficient to analyze the universality property for the three subcases where $\odot = \&$, $\odot = \vee$, and $\odot = \oplus$. Let us consider them one by one.

$7.1°$. For $\ominus = \vee$ and $\odot = \&$, the universality property is known to be true – this is one of the two basic cases with which we started our analysis.

$7.2°$. When $\ominus = \vee$ and $\odot = \vee$, the general expression

$$(A_1(x) \odot B_1(u)) \ominus (A_2(x) \odot B_2(u)) \ominus \dots$$

takes the form

$$(A_1(x) \vee B_1(u)) \vee (A_2(x) \vee B_2(u)) \vee \dots$$

Due to commutativity and associativity of the "or" operation, this expression is equivalent to

$$(A_1(x) \vee A_2(x) \vee \dots) \vee (B_1(u) \vee B_2(u) \vee \dots),$$

i.e., to $A(x) \vee B(u)$, where $A(x) \overset{\text{def}}{=} A_1(x) \vee A_2(x) \vee \dots$ and

$$B(u) \overset{\text{def}}{=} B_1(u) \vee B_2(u) \vee \dots.$$

Let us prove, by contradiction, that a property $(x = x_0) \& (u = u_0)$ cannot be represented as $A(x) \vee B(u)$:

$$((x = x_0) \& (u = u_0)) \Leftrightarrow (A(x) \vee B(u)). \tag{4.13}$$

Indeed, for $x = x_0$ and $u = u_1 \neq u_0$, the left-hand side of (4.13) is false, hence the right-hand side $A(x_0) \vee B(u_1)$ is false as well. Thus, both statements $A(x_0)$ and $B(u_1)$ are false.

Similarly, for $x = x_1 \neq x_0$ and $u = u_0$, both statements $A(x_1)$ and $B(u_0)$ are false. Since $A(x_0)$ and $B(u_0)$ are both false, the disjunction $A(x_0) \vee B(u_0)$ is also false, so due to (4.13), we would conclude that the statement

$$(x_0 = x_0) \& (u_0 = u_0)$$

is false, while in reality, this statement is true. The contradiction proves that the pair $(\vee, \vee)$ does not have the universality property.

$7.3°$. Let us prove that the subcase $\ominus = \vee$ and $\odot = \oplus$ does not lead to the universality property.

In this case, the general expression

$$(A_1(x) \odot B_1(u)) \ominus (A_2(x) \odot B_2(u)) \ominus \ldots$$

takes the form

$$(A_1(x) \oplus B_1(u)) \vee (A_2(x) \oplus B_2(u)) \vee \ldots \tag{4.14}$$

Let us prove, by contradiction, that for every $x_0 \in X$ and $u_0 \in U$, the property $C(x,u) \Leftrightarrow (x = x_0 \,\&\, u = u_0)$ cannot be represented in the form (4.14). To prove it, let us assume that this property can be represented in this form, for some properties $A_i(x)$ and $B_i(u)$, and let us show that this assumption leads to a contradiction.

For the above property $C(x,u)$, the set S of all the values for which this property is true consists of a single pair (x_0, u_0). Due to equivalence, this same set S is also the set of all the pairs for which the formula (4.14) holds.

Due to the known properties of the "or" operations, the set S of all the values (x, u) for which the formula (4.14) holds is equal to the union of the sets $S_i = \{(x,u) : A_i(x) \oplus B_i(u)\}$. Thus, each of the sets S_i is a subset of the set S: $S_i \subseteq S$. By our construction, the set S consists of only one element; thus, it has only two subsets: itself and the empty set. If all the sets S_i coincided with the empty set, then their intersection would also be equal to the empty set $\emptyset$, but it is equal to $S \neq \emptyset$. Thus, at least for one i, we have $S_i = S$. For this i, we have the equivalence

$$((x = x_0) \,\&\, (u = u_0)) \Leftrightarrow (A_i(x) \oplus B_i(u)). \tag{4.15}$$

Similarly to Part 6.3 of this proof, we can reduce this equivalence to the case where $A_i(x_0)$ is true, i.e., where $A_i(x_0) = 1$. So, in the remaining part of this subsection, we assume that $A_i(x_0) = 1$.

Now, for $x = x_0$ and $u = u_0$, the left-hand side of the equivalence (4.15) is true; hence the right-hand side $A_i(x_0) \oplus B_i(u_0)$ is true as well. Since we assumed that $A_i(x_0) = 1$, by the properties of "exclusive or", we thus conclude that $B_i(u_0) = 0$.

For $x = x_1 \neq x_0$ and $u = u_0$, the left-hand side of the equivalence (4.15) is false, hence the right-hand side $A_i(x_1) \oplus B_i(u_0)$ is false as well. Since, as we have already proven, $B_i(u_0)$ is false, we conclude that $A_i(x_1)$ is false.

Similarly, for $x = x_0$ and $u = u_1 \neq u_0$, the left-hand side of the equivalence (4.15) is false, hence the right-hand side $A_i(x_0) \oplus B_i(u_1)$ is false as well. Since $A_i(x_0)$ is true, we conclude that $B_i(u_1)$ is true.

Now, for $x = x_1$ and $u = u_1$, $A_i(x_1)$ is true and $B_i(u_1)$ are false, hence their combination $A_i(x_1) \oplus B_i(u_1)$ is true. So, due to (4.15), we would conclude that the statement $(x_1 = x_0) \,\&\, (u_1 = u_0)$ is true, but this statement is actually false. The contradiction proves that the representation (4.15) is indeed impossible, and so, the pair $(\oplus, \vee)$ does not have the universality property.

$8°$. The last case is where $\ominus = \oplus$. Similarly to the previous two cases, we analyze the three subcases where $\odot = \&$, $\odot = \vee$, and $\odot = \oplus$ one by one.

$8.1°$. The following explicit formula enables us to show that the pair $(\&, \oplus)$ has the universality property:

$$C(x,u) \Leftrightarrow \oplus_{(x_0,u_0) \in C}((x = x_0) \,\&\, (u = u_0)). \tag{4.16}$$

Indeed, we know that a similar formula (4.5) holds with "or" instead of "exclusive or". Here, the properties $(x = x_0)$ & $(u = u_0)$ corresponding to different pairs (x_0, u_0) are mutually exclusive, and thus, for these properties, "or" coincides with "exclusive or".

8.2°. To prove that the pair $(\vee, \oplus)$ has the universality property, we need the following auxiliary result:

$$((x = x_0) \& (u = u_0)) \Leftrightarrow ((x = x_0) \vee (u = u_0)) \oplus (x = x_0) \oplus (u = u_0). \qquad (4.17)$$

Indeed, this can be proven by considering all four possible cases: $x = x_0$ and $u = u_0$, $x = x_0$ and $u \neq u_0$, $x \neq x_0$ and $u = u_0$, $x \neq x_0$ and $u \neq u_0$. Thus, the expression (4.10) can be reformulated in the following equivalent form:

$$C(x, u) \Leftrightarrow \oplus_{(x_0, u_0) \in C}(((x = x_0) \vee (u = u_0)) \oplus (x = x_0) \oplus (u = u_0)).$$

Hence, the pair $(\vee, \oplus)$ indeed has the universality property.

8.3°. When $\ominus = \oplus$ and $\odot = \oplus$, the general expression

$$(A_1(x) \odot B_1(u)) \ominus (A_2(x) \odot B_2(u)) \ominus \ldots$$

takes the form

$$(A_1(x) \oplus B_1(u)) \oplus (A_2(x) \oplus B_2(u)) \oplus \ldots$$

Due to commutativity and associativity of the "exclusive or" operation, this expression is equivalent to

$$(A_1(x) \oplus A_2(x) \oplus \ldots) \oplus (B_1(u) \oplus B_2(u) \oplus \ldots),$$

i.e., to $A(x) \oplus B(u)$, where $A(x) \overset{\text{def}}{=} A_1(x) \oplus A_2(x) \oplus \ldots$ and

$$B(u) \overset{\text{def}}{=} B_1(u) \oplus B_2(u) \oplus \ldots.$$

We have already shown, in Parts 6.3 and 7.3 of this proof, that not every property $C(x, u)$ can be represented as $A(x) \oplus B(u)$: for example, the property $(x = x_0)$ & $(u = u_0)$ cannot be thus represented. So, the pair $(\oplus, \oplus)$ does not have the universality property.

The theorem is proven.

4.3 Selecting the Best Strategy within a Given Class: Use of Approximate Symmetries (and Similarities)

Need for Fuzzy "EXCLUSIVE OR" Operations. In the previous section, we have shown that, in addition to usual classes of strategies for intelligent control, classes that use "and" and "or" operations, we also need to consider "exclusive or" operations.

The need to consider exclusive or operations goes beyond intelligent control. Indeed, one of the main objectives of fuzzy logic is to formalize commonsense and expert reasoning. In commonsense and expert reasoning, people use logical connectives like "and" and "or". Depending on the context, commonsense "or" can mean both "inclusive or" – when "*A* or *B*" means that it is also possible to have both *A* and *B*, and "exclusive or" – when "*A* or *B*" means that one of the statements holds but not both.

For example, for a dollar, a vending machine can produce either a coke or a diet coke, but not both.

In mathematics and computer science, "inclusive or" is the one most frequently used as a basic operation. Because of this, fuzzy logic – an extension of usual logic to fuzzy statements characterized by "degree" of truth – is also mainly using "inclusive or" operations. However, since "exclusive or" is also used in commonsense and expert reasoning, there is a practical need for a fuzzy versions of this operation.

Comment. The "exclusive or" operation is actively used in computer design: since it corresponds to the bit-by-bit addition of binary numbers (the carry is the "and"). It is also actively used in quantum computing algorithms; see, e.g., [122].

Fuzzy versions of "exclusive or" operations are also known; see, e.g., [8]. These fuzzy versions are actively used in machine learning; see, e.g., [22, 84, 95, 129]. In particular, some of these papers (especially [95]) use a natural extension of fuzzy "exclusive or" from a binary to a k-ary operation.

A Crisp "EXCLUSIVE OR" Operation: A Reminder. As usual with fuzzy operations, the fuzzy "exclusive or" operation must be an extension of the corresponding crisp operation. In the traditional 2-valued logic, with two possible truth values 0 (false) and 1 (true), the "exclusive or" operation $\oplus$ is defined as follows: $0 \oplus 0 = 1 \oplus 1 = 0$ and $0 \oplus 1 = 1 \oplus 0 = 1$. Thus, the desired fuzzy "exclusive or" operation $f_\oplus(a,b)$ must satisfy the same properties:

$$f_\oplus(0,0) = f_\oplus(1,1) = 0; \quad f_\oplus(0,1) = f_\oplus(1,0) = 1. \tag{4.18}$$

Need for the Least Sensitivity: Reminder. Fuzzy logic operations deal with experts' degrees of certainty in their statements. These degrees are not precisely defined, the same expert can assign, say, 0.7 and 0.8 to the same degrees of belief. It is therefore reasonable to require that the result of the fuzzy operation does not change much if we slightly change the inputs. A reasonable way to formalize this "approximate symmetry" requirement is to require that the operation $f(a,b)$ satisfies the following property:

$$|f(a,b) - f(a',b')| \le k \cdot \max(|a - a'|, |b - b'|), \tag{4.19}$$

with the smallest possible value k among all operations $f(a,b)$ satisfying the given properties. Such operations are called *the least sensitive* or *the most robust*.

For T-Norms and T-Conorms, the Least Sensitivity Requirement Leads to Reasonable Operations. It is known that there is only one least sensitive t-norm ("and"-operation) $f_\&(a,b) = \min(a,b)$, and only one least sensitive t-conorm ("or"-operation) $f_\vee(a,b) = \max(a,b)$; see, e.g., [120, 116, 117].

What We Do in This Section. In this section, we describe the least sensitive fuzzy "exclusive or" operation.

Comment. This result first appeared in [55].

Definition 4.3.1. *A function* $f : [0,1] \times [0,1] \to [0,1]$ *is called a* fuzzy "exclusive or" *operation if it satisfies the following conditions:* $f(0,0) = f(1,1) = 0$ *and* $f(0,1) = f(1,0) = 1$.

Comment. We could also require other conditions, e.g., commutativity and associativity. However, our main objective is to select a single operation which is the least sensitive. The weaker the condition, the larger the class of operations that satisfy these conditions, and thus, the stronger the result that our operation is the least sensitive in this class.

Thus, to make our result as strong as possible, we selected the weakest possible condition – and thus, the largest possible class of "exclusive or" operations.

Definition 4.3.2. *Let F be a class of functions from* $[0,1] \times [0,1]$ *to* $[0,1]$. *We say that a function* $f \in F$ *is* the least sensitive *in the class F if for some real number k, the function f satisfies the condition*

$$|f(a,b) - f(a',b')| \leq k \cdot \max(|a-a'|, |b-b'|),$$

and no other function $f \in F$ *satisfies this condition.*

Theorem 4.3.1. *In the class of all fuzzy "exclusive or" operations, the following function is the least sensitive:*

$$f_\oplus(a,b) = \min(\max(a,b), \max(1-a, 1-b)). \tag{4.20}$$

Comments.

- This operation can be understood as follows. In the crisp (two-valued) logic, "exclusive or" $\oplus$ can be described in terms of the "inclusive or" operation $\vee$ as

$$a \oplus b \Leftrightarrow (a \vee b) \,\&\, \neg(a \,\&\, b).$$

If we:

  – replace $\vee$ with the least sensitive "or"-operation $f_\vee(a,b) = \max(a,b)$,
  – replace $\&$ with the least sensitive "and"-operation $f_\&(a,b) = \min(a,b)$, and
  – replace $\neg$ with the least sensitive negation operation $f_\neg(a) = 1-a$,

then we get the expression (4.20) given in the Theorem.
- The above operation is associative and has a value a_0 (equal to 0.5) which satisfies the property $a \oplus a_0 = a$ for all a. Thus, from the mathematical viewpoint, this operation is an example of a *nullnorm*; see, e.g., [9].

Proof of Theorem 4.3.1

We prove that the Theorem is true for $k = 1$.

$1°$. First, let us prove that the operation (4.20) indeed satisfies the condition (4.19) with $k = 1$. In other words, let us prove that for every $\varepsilon > 0$, if $|a - a'| \leq \varepsilon$ and $|b - b'| \leq \varepsilon$, then $|f_\oplus(a,b) - f_\oplus(a',b')| \leq \varepsilon$.

$1.1°$. It is known (see, e.g., [116, 117, 120]) that the functions $\min(a,b)$, $\max(a,b)$, and $1 - a$ satisfy the condition (4.19) with $k = 1$. In particular, this means that if $|a - a'| \leq \varepsilon$ and $|b - b'| \leq \varepsilon$, then we have

$$|\max(a,b) - \max(a',b')| \leq \varepsilon \tag{4.21}$$

and also

$$|(1 - a) - (1 - a')| \leq \varepsilon \text{ and } |(1 - b) - (1 - b')| \leq \varepsilon. \tag{4.22}$$

$1.2°$. From (4.22), by using the property (4.19) for the max operation, we conclude that

$$|\max(1 - a, 1 - b) - \max(1 - a', 1 - b')| \leq \varepsilon. \tag{4.23}$$

$1.3°$. Now, from (4.21) and (4.23), by using the property (4.19) for the min operation, we conclude that

$$|D| \leq \varepsilon, \tag{4.24}$$

where

$$D \stackrel{\text{def}}{=} \min(\max(a,b), \max(1 - a, 1 - b)) - \min(\max(a',b'), \max(1 - a', 1 - b')).$$

The statement is proven.

$2°$. Let us now assume that $f(a,b)$ is an exclusive or operation that satisfies the condition (4.19) with $k = 1$. Let us prove that then $f(a,b)$ coincides with the function (4.20).

$2.1°$. Let us first prove that $f(0.5, 0.5) = 0.5$.

The proof can be illustrated by the following picture.

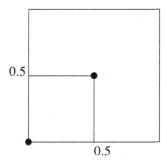

By the definition of the exclusive or operation, we have $f(0,0) = 0$ and $f(0,1) = 1$. Due to the property (4.19), we have

$$|f(0,0) - f(0.5,0.5)| \leq \max(|0 - 0.5|, |0 - 0.5|) = 0.5 \qquad (4.25)$$

thus,

$$f(0.5,0.5) \leq f(0,0) + 0.5 = 0 + 0.5 = 0.5. \qquad (4.26)$$

Similarly, due to the property (4.19), we have

$$|f(0,1) - f(0.5,0.5)| \leq \max(|0 - 0.5|, |1 - 0.5|) = 0.5 \qquad (4.27)$$

thus,

$$f(0.5,0.5) \geq f(0,1) - 0.5 = 1 - 0.5 = 0.5. \qquad (4.28)$$

From (4.26) and (4.28), we conclude that $f(0.5,0.5) = 0.5$.

$2.2°$. Let us now prove that $f(a,a) = a$ for $a \leq 0.5$.

This proof can be illustrated by the following picture.

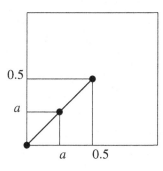

Due to the property (4.19), we have

$$|f(0,0) - f(a,a)| \leq \max(|0 - a|, |0 - a|) = a \qquad (4.29)$$

thus,

$$f(a,a) \leq f(0,0) + a = 0 + a = a. \tag{4.30}$$

Similarly, due to the property (4.19), we have

$$|f(0.5, 0.5) - f(a,a)| \leq \max(|0.5 - a|, |0.5 - a|) = 0.5 - a \tag{4.31}$$

thus,

$$f(a,a) \geq f(0.5, 0.5) - (0.5 - a) = 0.5 - (0.5 - a) = a. \tag{4.32}$$

From (4.30) and (4.32), we conclude that $f(a,a) = a$.

2.3°. Similarly:

- by considering the points $(0.5, 0.5)$ and $(1,1)$, we conclude that

$$f(1 - a, 1 - a) = a$$

for $a \leq 0.5$;

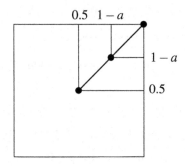

- by considering the points $(0.5, 0.5)$ and $(0,1)$, we conclude that

$$f(a, 1 - a) = 1 - a$$

for $a \leq 0.5$;

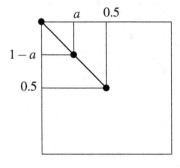

- by considering the points $(0.5, 0.5)$ and $(1, 0)$, we conclude that

$$f(1-a, a) = 1 - a$$

for $a \leq 0.5$.

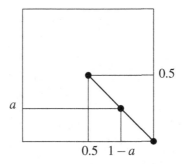

Summarizing: we have just proved that the formula (4.23) holds when $b = a$ and when $b = 1 - a$.

2.4°. Let us now prove that the formula (4.23) holds for arbitrary a and b.

In principle, we can have four cases depending on whether $b \leq a$ or $b \geq a$ and on whether $b \leq 1 - a$ or $b \geq 1 - a$. Without losing generality, let us consider the case where $b \leq a$ and $b \leq 1 - a$; the other three cases can be proven in a similar way.

The proof for this case can be illustrated by the following picture.

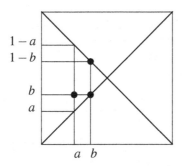

For this case, we know, from Parts 2.2 and 2.3 of this proof, that $f(b, b) = b$ and $f(1 - b, b) = 1 - b$. Here, $b \leq a \leq 1 - b$. Due to the property (4.19), we have

$$|f(a, b) - f(b, b)| \leq \max(|a - b|, |b - b|) = a - b, \tag{4.33}$$

thus,

$$f(a, b) \leq f(b, b) + (a - b) = b + (a - b) = a. \tag{4.34}$$

Similarly, due to the property (4.19), we have

$$|f(a,b) - f(1-b,b)| \leq \max(|a - (1-b)|, |b-b|) = (1-b) - a, \qquad (4.35)$$

thus,

$$f(a,b) \geq f(1-b,b) - ((1-b) - a) = (1-b) - ((1-b) - a) = a. \qquad (4.36)$$

From (4.34) and (4.36), we conclude that $f(a,b) = a$.

Similarly:

- for $b \leq a$ and $b \geq 1 - a$, i.e., when $1 - a \leq b \leq a$, by considering the points $(a, 1-a)$ and (a,a), we conclude that $f(a,b) = 1 - a$;
- for $b \geq a$ and $b \leq 1 - a$, i.e., when $a \leq b \leq 1 - a$, by considering the points (a,a) and $(a, 1-a)$, we conclude that $f(a,b) = b$;
- for $b \geq a$ and $b \geq 1 - a$, i.e., when $1 - b \leq a \leq b$, by considering the points $(1-b,b)$ and (b,b), we conclude that $f(a,b) = 1 - b$.

In other words, we prove that the formula (4.23) holds for all a and b. The theorem is proven.

Average Sensitivity: Reminder. As we have mentioned earlier, the fuzzy degrees are given with some uncertainty. In other words, different experts – and even the same expert at different times – would assign somewhat different numerical values to the same degree of certainty. In the main part of the section, we have showed how to select fuzzy operations $c = f(a,b)$ in such a way that "in the worst case", the change in a and b would lead to the smallest possible change in the value $c = f(a,b)$.

Another reasonable possibility is to select fuzzy operations $c = f(a,b)$ in such a way that "on average", the change in a and b would lead to the smallest possible change in the value $c = f(a,b)$.

For each pair of values a and b, it is reasonable to assume that the differences Δa and Δb between the different numerical values corresponding to the same degree of certainty are independent random variables with 0 mean and small variance σ^2. Since the differences Δa and Δb are small, we can expand the difference $\Delta c = f(a + \Delta a, b + \Delta b) - f(a,b)$ in Taylor series with respect to Δa and Δb and keep only linear terms in this expansion:

$$\Delta c \approx \frac{\partial f}{\partial a} \cdot \Delta a + \frac{\partial f}{\partial b} \cdot \Delta b. \qquad (4.37)$$

Since the variance is independent with 0 mean, the mean of Δc is also 0, and variance of Δc is equal to

$$\sigma^2(a,b) = \left(\left(\frac{\partial f}{\partial a} \right)^2 + \left(\frac{\partial f}{\partial b} \right)^2 \right) \cdot \sigma^2. \qquad (4.38)$$

This is the variance for given a and b. To get the average variance, it is reasonable to average this value over all possible values of a and b, i.e., to consider the value

$$I \cdot \sigma^2,$$

where

$$I \stackrel{\text{def}}{=} \int_{a=0}^{a=1} \int_{b=0}^{b=1} \left(\left(\frac{\partial f}{\partial a} \right)^2 + \left(\frac{\partial f}{\partial b} \right)^2 \right) da\,db. \qquad (4.39)$$

Thus, the average sensitivity is the smallest if, among all possible functions $f(a,b)$ satisfying the given constraints, we select a function for which the integral I takes the smallest possible value.

Average Sensitivity: Known Results. [118, 120]

- For negation operations, this approach selects the standard function

$$f_\neg(a) = 1 - a.$$

- For "and"-operations (t-norms), this approach selects $f_\&(a,b) = a \cdot b$.
- For "or"-operations (t-conorms), this approach selects $f_\vee(a,b) = a + b - a \cdot b$.

New Result: Formulation. We consider *"exclusive or"* operations, i.e., functions $f(a,b)$ from $[0,1] \times [0,1]$ to $[0,1]$ for which $f(0,b) = b$, $f(a,0) = a$, $f(1,b) = 1 - b$, and $f(a,1) = 1 - a$.

Our main result is that among all such operations, the operation which is the least sensitive on average has the form

$$f_\oplus(a,b) = a + b - 2 \cdot a \cdot b. \qquad (4.40)$$

Comment. This operation can be explained as follows:

- First, we represent the classical (2-valued) "exclusive or" operation $a \oplus b$ as $(a \vee b)\&(\neg a \vee \neg b)$.
- Then, to get a fuzzy analogue of this operation, we replace $p \vee q$ with $p + q - p \cdot q$, $\neg p$ with $1 - p$, and $p \& q$ with $\max(p + q - 1, 0)$.

Indeed, in this case,

$$a \vee b = a + b - a \cdot b;$$

$$\neg a \vee \neg b = (1 - a) \vee (1 - b) = (1 - a) + (1 - b) - (1 - a) \cdot (1 - b) =$$

$$1 - a + 1 - b - (1 - a - b + a \cdot b) = 1 - a + 1 - b - 1 + a + b - a \cdot b = 1 - a \cdot b,$$

and thus,

$$(a \vee b) + (\neg a \vee \neg b) - 1 = a + b - a \cdot b + 1 - a \cdot b - 1 = a + b - 2 \cdot a \cdot b.$$

For values $a, b \in [0,1]$, we have $a^2 \le a$ and $b^2 \le b$, hence

$$(a \vee b) + (\neg a \vee \neg b) - 1 = a + b - 2 \cdot a \cdot b \ge a^2 + b^2 - 2 \cdot a \cdot b = (a - b)^2 \ge 0,$$

therefore, indeed

$$(a \vee b)\&(\neg a \vee \neg b) = \max((a \vee b) + (\neg a \vee \neg b) - 1, 0) = (a \vee b) + (\neg a \vee \neg b) - 1.$$

This replacement operation sounds arbitrary, but the resulting "exclusive or" operation is uniquely determined by the sensitivity requirement.

Proof of the Auxiliary Result. It is known similarly to the fact that the minimum of a function is always attained at a point where its derivative is 0, the minimum of a functional is always attained at a function where its *variational derivative* is equal to 0 (see, e.g., [50]; see also [118, 120]):

$$\frac{\delta L}{\delta f} = \frac{\partial L}{\partial f} - \sum_i \frac{\partial}{\partial x_i}\left(\frac{\partial L}{\partial f_i}\right) = 0,$$

where $f_{,i} \overset{\text{def}}{=} \dfrac{\partial f}{\partial x_i}$.

Applying this *variational equation* to the functional $I = \int L\,da\,db$, with $L = \left(\dfrac{\partial f}{\partial a}\right)^2 + \left(\dfrac{\partial f}{\partial b}\right)^2$, we conclude that

$$-\frac{\partial}{\partial a}\left(2 \cdot \frac{\partial f}{\partial a}\right) - \frac{\partial}{\partial b}\left(2 \cdot \frac{\partial f}{\partial b}\right) = 0,$$

i.e., we arrive at the equation

$$\nabla^2 f = 0, \tag{4.41}$$

where $\nabla \overset{\text{def}}{=} \left(\dfrac{\partial f}{\partial a}, \dfrac{\partial f}{\partial b}\right)$ and

$$\nabla^2 f = \frac{\partial^2 f}{\partial a^2} + \frac{\partial^2 f}{\partial b^2}.$$

The equation (4.41) is known as the *Laplace equation*, and it is known (see, e.g., [36]) that a solution to this equation is uniquely determined by the boundary conditions – i.e., in our case, by the values on all four parts of the boundary of the square $[0,1] \times [0,1]$: lines segments $a = 0$, $a = 1$, $b = 0$, and $b = 1$. One can easily show that the above function $f(a,b) = a + b - 2 \cdot a \cdot b$ satisfies the Laplace equation – since both its second partial derivatives are simply 0s. It is also easy to check that for all four sides, this function coincides with our initial conditions:

- when $a = 0$, we get $f(a,b) = 0 + b - 2 \cdot 0 \cdot b = b$;
- when $a = 1$, we get $f(a,b) = 1 + b - 2 \cdot 1 \cdot b = 1 - b$;
- when $b = 0$, we get $f(a,b) = a + 0 - 2 \cdot a \cdot 0 = a$;
- when $b = 1$, we get $f(a,b) = a + 1 - 2 \cdot 1 \cdot a = 1 - a$.

Thus, due to the above property of the Laplace equation, the function $f(a,b) = a + b - 2 \cdot a \cdot b$ is the only solution to this equation with the given initial condition – therefore, it coincides with the desired *the least sensitive on average* "exclusive or" operation (which satisfies the same Laplace equation with the same boundary conditions).

The theorem is proven.

4.4   Selecting the Best Strategy within a Given Class: Optimality Naturally Leads to Symmetries (and Similarities)

Formulation of the Problem. In the previous section, we found the fuzzy logical operations which are the most invariant with respect to changes in degrees of certainty. In the derivation of these operations, we accepted the approach based on symmetry and similarity, and found the operations generated by this approach. In practice, the optimality criterion is often formulated in precise terms: e.g., we want to find the control which is the most stable or the smoothest. In this section, we show that such natural optimality criteria also lead to operations based on symmetry and similarity. The corresponding results provide an additional justification for the symmetry-and-similarity approach.

In order to describe our result, we need to describe fuzzy control methodology in detail.

Fuzzy Control Methodology: A Detailed Introduction. In the situations where we do not have the complete knowledge of the plant, we often have the experience of human operators who successfully control this plant. We would like to make an automated controller that uses their experience. With this goal in mind, an ideal situation is where an operator can describe his control strategy in precise mathematical terms. However, most frequently, the operators cannot do that (can you describe how exactly you drive your car?). Instead, they explain their control in terms of *rules* formulated in natural language (like "if the velocity is high, and the obstacle is close, break immediately"). *Fuzzy control* is a methodology that translates these natural-language rules into an automated control strategy. This methodology was first outlined by L. Zadeh [23] and experimentally tested by E. Mamdani [93] in the framework of *fuzzy* set theory [156] (hence the name). For many practical systems, this approach works fine.

Specifically, the rules that we start with are usually of the following type:

if x_1 is A_1^j and x_2 is A_2^j and ... and x_n is A_n^j, then u is B^j

where x_i are parameters that characterize the plant, u is the control, and A_i^j, B^j are the terms of natural language that are used in describing the j-th rule (e.g., "small", "medium", etc).

The value u is an appropriate value of the control if and only if at least one of these rules is applicable. Therefore, if we use the standard mathematical notations & for "and", $\vee$ for "or", and $\equiv$ for "if and only if", then the property "u is an appropriate control" (which we denote by $C(u)$) can be described by the following informal "formula":

$$C(u) \equiv (A_1^1(x_1) \& A_2^1(x_2) \& \dots \& A_n^1(x_n) \& B^1(u)) \vee$$

$$(A_1^2(x_1) \& A_2^2(x_2) \& \dots \& A_n^2(x_n) \& B^2(u)) \vee$$

$$\dots$$

$$(A_1^K(x_1) \& A_2^K(x_2) \& \dots \& A_n^K(x_n) \& B^K(u))$$

Terms of natural language are described as membership functions. In other words, we describe $A_i^j(x)$ as $\mu_{j,i}(x)$, the degree of belief that a given value x satisfies the property A_i^j. Similarly, $B^j(u)$ is represented as $\mu_j(u)$. Logical connectives & and $\vee$ are interpreted as some operations $f_\vee$ and $f_\&$ with degrees of belief (e.g., $f_\vee = \max$ and $f_\& = \min$). After these interpretations, we can form the membership function for control: $\mu_C(u) = f_\vee(p_1, \ldots, p_K)$, where

$$p_j = f_\&(\mu_{j,1}(x_1), \mu_{j,2}(x_2), \ldots, \mu_{j,n}(x_n), \mu_j(u))), \quad j = 1, \ldots, K.$$

We need an automated control, so we must end up with a single value $\bar{u}$ of the control that will actually be applied. An operation that transforms a membership function into a single value is called a *defuzzification*. Therefore, to complete the fuzzy control methodology, we must apply some defuzzification operator D to the membership function $\mu_C(u)$ and thus obtain the desired value $\bar{u} = f_C(\mathbf{x})$ of the control that corresponds to $\mathbf{x} = (x_1, \ldots, x_n)$. Usually, the *centroid defuzzification* is used, where

$$\bar{u} = \frac{\int u \cdot \mu_C(u)\, du}{\int \mu_C(u)\, du}.$$

A Simple Example: Controlling a Thermostat. The goal of a thermostat is to keep a temperature T equal to some fixed value T_0, or, in other words, to keep the difference $x = T - T_0$ equal to 0. To achieve this goal, one can control the degree of cooling or heating. What we actually control is the rate at which the temperature changes, i.e., in mathematical terms, a derivative $\dot{T}$ of temperature with respect to time. So if we apply the control u, the behavior of the thermostat is determined by the equation $\dot{T} = u$. In order to automate this control we must design a function $u(x)$ that describes what control to apply if the temperature difference x is known.

In many cases, the exact dependency of the temperature on the control is not precisely known. Instead, we can use our experience, and formulate reasonable control rules:

- If the temperature T is close to T_0, i.e., if the difference $x = T - T_0$ is negligible, then no control is needed, i.e., u is also negligible.
- If the room is slightly overheated, i.e., if x is positive and small, we must cool it a little bit (i.e., $u = \dot{x}$ must be negative and small).
- If the room is slightly overcooled, then we need to heat the room a little bit. In other terms, if x is small negative, then u must be small positive.

So, we have the following rules:

- if x is negligible, then u must be negligible;
- if x is small positive, then u must be small negative;
- if x is small negative, then u must be small positive.

In this case, u is a reasonable control if either:

- the first rule is applicable (i.e., x is negligible) and u is negligible; or
- the second rule is applicable (i.e., x is small positive), and u must be small negative;

- or the third rule is applicable (i.e., x is small negative), and u must be small positive.

Summarizing, we can say that u is an appropriate choice for a control if and only if either x is negligible and u is negligible, or x is small positive and u is small negative, etc. If we use the denotations $C(u)$ for "u is an appropriate control", $N(x)$ for "x is negligible", SP for "small positive", and SN for "small negative", then we arrive at the following informal "formula":

$$C(u) \equiv (N(x)\&N(u)) \vee (SP(x)\&SN(u)) \vee (SN(x)\&SP(u)).$$

If we denote the corresponding membership functions by μ_N, μ_{SP}, and μ_{SN}, then the resulting membership function for control is equal to

$$\mu_C(u) = f_\vee(f_\&(\mu_N(x), \mu_N(u)), f_\&(\mu_{SP}(x), \mu_{SN}(u)), f_\&(\mu_{SN}(x), \mu_{SP}(u))).$$

Problem. There exist several versions of fuzzy control methodology. The main difference between these versions is in how they translate logical connectives "or" and "and", i.e., in other words, what *reasoning method* a version uses. Which of these versions should we choose? The goal of this section is to provide an answer to this question.

The Contents of This Section. The main criterion for choosing a set of reasoning methods is to achieve the best control possible. So, before we start the description of our problem, it is necessary to explain when a control is good. This is done in this section, first informally, then formally. Once we know what our objective is, we must describe the possible choices, i.e., the possible reasoning methods.

We are going to prove several results explaining what choice of a reasoning method leads to a better control. The proofs are very general. However, for the readers' convenience, we explain them on the example of a simple plant. This simple plant serves as a testbed for different versions of fuzzy control.

The formulation of the problem in mathematical terms is now complete. Now, we formulate the results, and describe the proofs of these results.

What Do We Expect from an Ideal Control? In some cases, we have a well-defined control objective (e.g., minimizing fuel). But in most cases, engineers do not explain explicitly what exactly they mean by an *ideal* control. However, they often do not hesitate to say that one control is better than another one. What do they mean by that? Usually, they draw a graph that describes how an initial perturbation changes with time, and they say that a control is good if this perturbation quickly goes down to 0 and then stays there.

In other words, in a typical problem, an ideal control consists of two stages:

- On the *first stage*, the main objective is to make the difference $x = X - X_0$ between the actual state X of the plant and its ideal state X_0 go to 0 as fast as possible.

- After we have already achieved the objective of the first stage, and the difference is close to 0, then the *second stage* starts. On this second stage, the main objective is to keep this difference close to 0 at all times. We do not want this difference to oscillate wildly, we want the dependency $x(t)$ to be as smooth as possible.

This description enables us to formulate the objectives of each stage in precise mathematical terms.

First Stage of the Ideal Control: Main Objective. We have already mentioned in Section 1 that, for readers' convenience, we illustrate our ideas on some simple plants. So, let us consider the case where the state of the plant is described by a single variable x, and we control the first time derivative $\dot{x}$. For this case, we arrive at the following definition:

Definition 4.4.1. *Let a function $u(x)$ be given; this function is called a* control strategy.

- *By a* trajectory *of the plant, we understand the solution of the differential equation $\dot{x} = u(x)$.*
- *Let's fix some positive number M (e.g., $M = 1000$). Assume also that a real number $\delta \neq 0$ is given. This number is called an* initial perturbation.
- *A relaxation time $t(\delta)$ for the control $u(x)$ and the initial perturbation δ is defined as follows:*
  - *we find a trajectory $x(t)$ of the plant with the initial condition $x(0) = \delta$, and*
  - *take as $t(\delta)$, the first moment of time starting from which $|x(t)| \leq |x(0)|/M$ (i.e., for which this inequality is true for all $t \geq t(\delta)$).*

Comment. For *linear control*, i.e., when $u(x) = -k \cdot x$ for some constant k, we have $x(t) = x(0) \exp(-k \cdot t)$ and therefore, the relaxation time t is easily determined by the equation $\exp(-k \cdot t) = 1/M$, i.e., $t = \ln(M/k)$. Thus defined relaxation time does not depend on δ. So, for control strategies that use linear control on the first stage, we can easily formulate the objective: to minimize relaxation time. The smaller the relaxation time, the closer our control to the ideal.

In the *general case*, we would also like to minimize relaxation time. However, in general, we encounter the following problem: For *non-linear control* (and fuzzy control *is* non-linear) the relaxation time $t(\delta)$ depends on δ. If we pick a δ and minimize $t(\delta)$, then we get good relaxation for this particular δ, but possibly at the expense of not-so-ideal behavior for different values of the initial perturbation δ.

How can we solve our problem? The problem that we encountered was due to the fact that we considered a simplified control situation, where we start to control a system only when it is already out of control. This may be too late. Usually, no matter how smart the control is, if a perturbation is large enough, the plant will never stabilize. For example, if the currents that go through an electronic system exceed a certain level, they simply burn the electronic components. To avoid that, we usually control the plant from the very beginning, thus preventing the values of x from becoming too large. From this viewpoint, what matters is how fast we go down for *small* perturbations, when $\delta \approx 0$.

What does "small" mean in this definition? If for some value δ that we initially thought to be small, we do not get a good relaxation time, then we try to keep the perturbations below that level. On the other hand, the smaller the interval that we want to keep the system in, the more complicated and costly this control becomes. So, we would not decrease the admissible level of perturbations unless we get a really big increase in relaxation time. In other words, we decrease this level (say, from δ_0 to $\delta_1 < \delta_0$) only if going from $t(\delta_0)$ to $t(\delta_1)$ means decreasing the relaxation time. As soon as $t(\delta_1) \approx t(\delta_0)$ for all $\delta_1 < \delta_0$, we can use δ_0 as a reasonable upper level for perturbations.

In mathematical terms, this condition means that $t(\delta_0)$ is close to the limit of $t(\delta)$ when $\delta \to 0$. So, the smaller this limit, the faster the system relaxes. Therefore, this limit can be viewed as a reasonable objective for the first stage of the control.

Definition 4.4.2. *By a* relaxation time *T for a control $u(x)$, we mean the limit of $t(\delta)$ for $\delta \to 0$.*

So, *the main objective of the first stage of control is to maximize relaxation time.*

Lemma 4.4.1. *If the control strategy $u(x)$ is a smooth function of x, then the relaxation time equals to $\ln M/(-u'(0))$, where u' denotes the derivative of u.*

Comment. So the bigger this derivative, the smaller the relaxation time. Therefore, our objective can be reformulated as follows: *to maximize $u'(0)$.*

Second Stage of the Ideal Control: Main Objective. After we have made the difference x go close to 0, the second stage starts, on which $x(t)$ has to be kept as smooth as possible. What does *smooth* mean in mathematical terms? Usually, we say that a trajectory $x(t)$ is smooth at a given moment of time t_0 if the value of the time derivative $\dot{x}(t_0)$ is close to 0. We want to say that a trajectory is smooth if $\dot{x}(t)$ is close to 0 for all t.

In other words, if we are looking for a control that is the smoothest possible, then we must find the control strategy for which $\dot{x}(t) \approx 0$ for all t. There are infinitely many moments of time, so even if we restrict ourselves to control strategies that depend on finitely many parameters, we have infinitely many equations to determine these parameters. In other words, we have an *over-determined* system. Such situations are well-known in data processing, where we often have to find parameters $p_1, \ldots, p_n$ from an over-determined system $f_i(p_1, \ldots, p_n) \approx q_i, 1 \le i \le N$. A well-known way to handle such situations is to use the *least squares method*, i.e., to find the values of p_j for which the "average" deviation between f_i and q_i is the smallest possible. To be more precise, we minimize the sum of the squares of the deviations, i.e., we are solving the following minimization problem:

$$\sum_{i=1}^{N} (f_i(p_1, \ldots, p_n) - q_i)^2 \to \min_{p_1, \ldots, p_n}.$$

In our case, $f_i = \dot{x}(t)$ for different moments of time t, and $q_i = 0$. So, least squares method leads to the criterion $\sum (\dot{x}(t))^2 \to \min$. Since there are infinitely many moments of time, the sum turns into an integral, and the criterion for choosing a control into $J(x(t)) \to \min$, where $J(x(t)) = \int (\dot{x}(t))^2 dt$. This value J thus represents a degree to which a given trajectory $x(t)$ is non-smooth. So, we arrive at the following definition:

Definition 4.4.3. *Assume that a control strategy $x(t)$ is given, and an initial perturbation δ is given. By a* non-smoothness $I(\delta)$ *of a resulting trajectory $x(t)$, we understand the value $J(x) = \int_0^\infty (\dot{x}(t))^2 dt$.*

Foundational comment. The least squares method is not only heuristic, it has several reasonable justifications. So, instead of simply borrowing the known methodology from data processing (as we did), we can formulate reasonable conditions for a functional J (that describes non-smoothness), and thus deduce the above-described form of J without using analogies at all. This is done in [79].

Mathematical Comment. What control to choose on the second stage? Similarly to relaxation time, we get different criteria for choosing a control if we use values of non-smoothness that correspond to different δ. And similarly to relaxation time, a reasonable solution to this problem is to choose a control strategy for which in the limit $\delta \to 0$, the non-smoothness takes the smallest possible value.

Mathematically, this solution is a little bit more difficult to implement than the solution for the first stage: Indeed, the relaxation time $t(\delta)$ has a well-defined non-zero limit when $\delta \to 0$, while non-smoothness simply tends to 0. Actually, for linear control, $I(\delta)$ tends to 0 as δ^2. To overcome this difficulty and still get a meaningful limit of non-smoothness, we divide $J(x)$ (and, correspondingly, $I(\delta)$) by δ^2 and only then, tend this ratio $\tilde{J}(x(t)) = \tilde{I}(\delta)$ to a limit. This division does not change the relationship between the functional and smoothness: indeed, if for some δ, a trajectory $x_1(t)$ is smoother than a trajectory $x_2(t)$ in the sense that $J(x_1(t)) < J(x_2(t))$, then, after dividing both sides by δ^2, we get $\tilde{J}(x_1(t)) < \tilde{J}(x_2(t))$. So, a trajectory $x(t)$ for which $\tilde{J}(x)$ is smaller, is thus smoother.

As a result, we arrive at the following definition.

Definition 4.4.4. *By a* non-smoothness I *of a control $u(x)$, we mean the limit of $I(\delta)/\delta^2$ for $\delta \to 0$.*

So, *the main objective of the second stage of control is to minimize non-smoothness.*

General Properties of "OR"- and "AND"-Operations: Commutativity and Associativity. In order to apply fuzzy control methodology, we must assign a truth value (also called *degree of belief*, or *certainty value*) $t(A)$ to every uncertain statement A contained in the experts' rules. Then, we must define "or"- and "and"-operations $f_\vee(a, b)$ and $f_\&(a, b)$ in such a way that for generic statements A and B, $t(A \vee B)$ is close to $f_\vee(t(A), t(B))$, and $t(A\&B)$ is close to $f_\&(t(A), t(B))$. Let us first describe properties that are general to both "or"- and "and"-operations.

Statements $A\&B$ and $B\&A$ mean the same. Hence, $t(A\&B) = t(B\&A)$, and it is therefore reasonable to expect that $f_\&(t(A), t(B)) = f_\&(t(B), t(A))$ for all A and B. In other words, it is reasonable to demand that $f_\&(a, b) = f_\&(b, a)$ for all a and b, i.e., that $f_\&$ is a *commutative* operation. Similarly, it is reasonable to demand that $f_\vee$ is a commutative operation.

Statements $(A\&B)\&C$ and $A\&(B\&C)$ also mean the same thing: that all three statements A, B, and C are true. Therefore, it is reasonable to demand that the corresponding approximations $f_\&(f_\&(t(A),t(B)),t(C))$ and $f_\&(t(A),f_\&(t(B),t(C)))$ coincide. In mathematical terms, it means that an "and"-operation must be *associative*. Similarly, it is reasonable to demand that an "or"-operation is associative. To make our exposition complete, let us give a precise mathematical definition.

Definition 4.4.5. *A function* $f : [0,1] \times [0,1] \to [0,1]$ *is called* commutative *if* $f(a,b) = f(b,a)$ *for all a and b. It is called* associative *if* $f(f(a,b),c) = f(a,f(b,c))$ *for all a, b, c.*

Comment. If a function f is commutative and associative, then the result of applying f to several values $a,b,\ldots,c$ does not depend on their order. So, we can use a simplified notation $f(a,b,\ldots,c)$ for $f(a,f(b,\ldots c)\ldots))$.

What are the possible "or"-operations? One of the most frequently used methods of assigning a certainty value $t(A)$ to a statement A is as follows (see, e.g., [14, 15]; [33], IV.1.d; [69]): we take several (N) experts, and ask each of them whether he believes that a given statement A is true (for example, whether he believes that 0.3 is negligible). If $N(A)$ of them answer "yes", we take the ratio $t(A) = N(A)/N$ as a desired certainty value. In other words, we take $t(A) = |S(A)|/N$, where $S(A)$ is the set of all experts (out of the given N) who believe that A is true, and $|S|$ denotes the number of elements in a given set S. Here, $S(A \vee B) = S(A) \cup S(B)$, hence,

$$N(A \vee B) = |S(A \cup B)| \leq |S(A)| + |S(B)| = N(A) + N(B).$$

If we divide both sides of this inequality by N, we can conclude that $t(A \vee B) \leq t(A) + t(B)$. Also, since $N(A) \leq N$, we get $t(A) \leq 1$, hence,

$$t(A \vee B) \leq \min(t(A) + t(B), 1).$$

On the other hand, since $S(A) \subseteq S(A) \cup S(B)$, we have $|S(A)| \leq |S(A \vee B)|$ and hence, $t(A) \leq t(A \vee B)$. Similarly, $t(B) \leq t(A \vee B)$. From these two inequalities, we can deduce that $\max(t(A),t(B)) \leq t(A \vee B)$. So, we arrive at the following definition:

Definition 4.4.6. *By an* "or"-operation*, we understand a commutative and associative function* $f_\vee : [0,1] \times [0,1] \to [0,1]$ *for which*

$$\max(a,b) \leq f_\vee(a,b) \leq \min(a+b,1)$$

for all a and b.

Comment. Another possibility to estimate $t(A)$ is to interview a single expert and express his degree of confidence in terms of the so-called *subjective probabilities* [142]. For this method, similar inequalities can be extracted from the known properties of (subjective) probabilities.

What Are the Possible "AND"-Operations? Similarly to $\vee$, we can conclude that $S(A\&B) = S(A) \cap S(B)$, so $N(A\&B) \leq N(A)$, $N(A\&B) \leq N(B)$, hence $N(A\&B) \leq \min(N(A),N(B))$ and

$$t(A\&B) \leq \min(t(A), t(B)).$$

On the other hand, a person does not believe in $A\&B$ iff either he does not believe in A, or he does not believe in B. Therefore, the number $N(\neg(A\&B))$ of experts who do not believe in $A\&B$ cannot exceed the sum $N(\neg A) + N(\neg B)$. The number $N(\neg(A\&B))$ of experts who do not believe in $A\&B$ is equal to $N - N(A\&B)$, and similarly, $N(\neg A) = N - N(A)$ and $N(\neg B) = N - N(B)$. Therefore, the above-mentioned inequality turns into

$$N - N(A\&B) \leq N - N(A) + N - N(B),$$

which leads to $N(A\&B) \geq N(A) + N(B) - N$ and hence, to $t(A\&B) \geq t(A) + t(B) - 1$. Since $t(A\&B) \geq 0$, we have

$$t(A\&B) \geq \max(0, t(A) + t(B) - 1).$$

So, we arrive at the following definition:

Definition 4.4.7. *By an* "and"*-operation, we understand a commutative and associative function* $f_\& : [0,1] \times [0,1] \to [0,1]$ *for which*

$$\max(0, a + b - 1) \leq f_\&(a,b) \leq \min(a,b)$$

for all a and b.

Comment. The same formulas hold if we determine $t(A)$ as a subjective probability.

Problems with "AND"-Operations. The definition that we came up with for an "or"-operation was OK, but with "and"-operations, we have a problem: in some situations, an "and"-operation can be unusable for fuzzy control. For example, if $f_\&(a,b) = 0$ for some $a > 0$, $b > 0$, then for some $x, \dot{x}, \ldots$ the resulting membership function for a control $\mu_C(u)$ can be identically 0, and there is no way to extract a value of the control $\bar{u}$ from such a function. For such situations, it is necessary to further restrict the class of possible "and"-operations.

In the following subsection, we describe how this problems can be solved.

Solution to the Problem: Correlated "AND"-Operations. We have already mentioned that to solve the first problem (that $\mu_C(u)$ is identically 0 and hence, no fuzzy control is defined), we must restrict the class of possible "and"-operations. The forthcoming restriction is based on the following idea. If belief in A and belief in B were independent events (in the usual statistical sense of the word "independent"), then we would have $t(A\&B) = t(A) \cdot t(B)$. In real life, beliefs are not independent. Indeed, if an expert has strong beliefs in several statements that later turn out to be true, then this means that he is really a good expert. Therefore, it is reasonable to expect that his degree of belief in other statements that are actually true is bigger than the degree of belief of an average expert. If A and B are statements with $t(A) > 1/2$ and $t(B) > 1/2$, i.e., such that the majority of experts believe in A and in B, this means that there is a huge possibility that both A and B are actually true. A reasonable portion of the experts are *good experts*, i.e., experts whose predictions are almost often true. All of these good experts believe in A and in B and therefore, all of them believe in $A\&B$.

Let us give an (idealized) numerical example of this phenomenon. Suppose that, say, 60% of experts are good, and $t(A) = t(B) = 0.7$. This means that at least some of these good experts believe in A, and some believe in B. Since we assumed that the beliefs of good experts usually come out right, it means that A and B are actually true. Therefore, because of the same assumption about good experts, all good experts believe in A, and all good experts believe in B. Therefore, all of them believe in $A\&B$. Hence,

$$t(A\&B) \geq 0.6 > t(A) \cdot t(B) = 0.49.$$

In general, we have a mechanism that insures that there is, in statistical terms, a positive *correlation* between beliefs in A and B. In mathematical terms, the total number $N(A\&B)$ of experts who believe in $A\&B$ must be larger than the number $N_{ind}(A\&B) = Nt(A)t(B) = N(N(A)/N)(N(B)/N)$ that corresponds to the case where beliefs in A and B are uncorrelated random events. So we come to a conclusion that the following inequality sounds reasonable: $t(A\&B) \geq t(A) \cdot t(B)$. So, we arrive at the following definition:

Definition 4.4.8. *An "and"-operation is called* correlated *if $f_\&(a,b) \geq a \cdot b$ for all a,b.*

Comment. In this case, we are guaranteed that if $a > 0$ and $b > 0$, then $f_\&(a,b) > 0$, i.e., we do avoid the problem in question.

Let's Describe a Simplified Plant, on Which Different Reasoning Methods Are Tested. We consider the simplest case where the state of the plant is described by a single variable x, and we control the first time derivative $\dot{x}$. To complete our description of the control problem, we must also describe:

- the experts' *rules*,
- the corresponding *membership functions*, and
- defuzzification.

Membership Functions. For simplicity, we consider the simplest (and most frequently used; see, e.g., [80, 81, 82]) membership functions, namely, triangular ones (as we can see from our proof, the result does not change if we use any other type of membership functions).

Definition 4.4.9. *By a triangular membership function with a midpoint a and endpoints $a - \Delta_1$ and $a + \Delta_2$ we mean the following function $\mu(x)$:*

- $\mu(x) = 0$ *if $x < a - \Delta_1$ or $x > a + \Delta_2$;*
- $\mu(x) = (x - (a - \Delta_1))/\Delta_1$ *if $a - \Delta_1 \leq x \leq a$;*
- $\mu(x) = 1 - (x - a)/\Delta_2$ *if $a \leq x \leq a + \Delta_2$.*

Rules. Fuzzy control can be viewed as a kind of extrapolation. In reality there exists some control $u(x,\ldots)$ that an expert actually applies. However, he cannot precisely explain, what function u he uses. So we ask him lots of questions, extract several rules, and form a fuzzy control from these rules.

We restrict ourselves to the functions $u(x)$ that satisfy the following properties:

Definition 4.4.10. *By an* actual control function *(or* control function, *for short), we mean a function $u(x)$ that satisfies the following three properties:*

- $u(0) = 0$;
- $u(x)$ *is monotonically decreasing for all x;*
- $u(x)$ *is smooth (differentiable).*

Comment. These restrictions are prompted by common sense:

- If $x = 0$, this means that we are already in the desired state, and there is no need for any control, i.e., $u(0) = 0$.
- The more we deviate from the desired state $x = 0$, the faster we need to move back if we want the plant to be controllable. So, u is monotonically decreasing.
- We want the control to be smooth (at least on the second stage), so the function $u(x)$ that describes an expert's control, must be smooth.

Let's now describe the resulting rules formally.

Definition 4.4.11. *Let's fix some $\Delta > 0$. For every integer j, by N_j, we denote a triangular membership function with a midpoint $j \cdot \Delta$ and endpoints $(j-1) \cdot \Delta$ and $(j+1) \cdot \Delta$.*

- *We call the corresponding fuzzy property N_0 negligible (N for short), N_1 small positive or SP, and N_{-1} small negative, or SN.*
- *Assume that a monotonically non-increasing function $u(x)$ is given, and that $u(0) = 0$. By rules generated by $u(x)$, we mean the set of following rules: "if $N_j(x)$, then $M_j(u)$" for all u, where M_j is a triangular membership function with a midpoint $u(j \cdot \Delta)$ and endpoints $u((j-1) \cdot \Delta)$ and $u((j+1) \cdot \Delta)$.*

In particular, if we start with a linear control $u = -k \cdot x$ (and linear control is the one that is most frequently used, see. e.g., [26]), then M_j resembles N_{-j} with the only difference being that instead of Δ, we use $k\Delta$. So, we can reformulate the corresponding rules as follows: if x is negligible, then u must be negligible; if x is small positive, then u must be small negative, etc. Here, we use Δ when we talk about x, and we use $k\Delta$ when we talk about u.

How to choose Δ? We have two phenomena to take into consideration:

- On one hand, the smaller Δ, the better the resulting rules represent the original expert's control. From this viewpoint, the smaller Δ, the better.
- On the the other hand, the smaller Δ, the more rules we have and therefore, the more running time our control algorithm requires. So, we must not take Δ too small.

As a result, the following is the natural way to choose Δ:

- choose some reasonable value of Δ;

- if the resulting control is not good enough, decrease Δ;
- repeat this procedure until the further decrease does not lead to any improvement in the control quality.

So, the quality (i.e., relaxation time or non-smoothness) of the rule-based control for the chosen Δ is close to the limit value of this quality when $\Delta \rightarrow 0$. Therefore, when choosing the best reasoning method, we must consider this limit quality as a choosing criterion. Let's formulate the relevant definitions.

Definition 4.4.12. *Assume that the following are given:*

- *an actual control function $u(x)$;*
- *a defuzzification procedure.*

For a given $\Delta > 0$, by a Δ-relaxation time, we mean the relaxation time of a control strategy that is generated by an actual control function $u(x)$ for this Δ. By a relaxation time, corresponding to an actual control function $u(x)$, *we mean the limit of Δ-relaxation times when $\Delta \rightarrow 0$.*

Definition 4.4.13. *Assume that the following are given:*

- *an actual control function $u(x)$;*
- *a defuzzification procedure.*

For a given $\Delta > 0$, by a Δ-non-smoothness, we mean the non-smoothness of a control strategy that is generated by an actual control function $u(x)$ for this Δ. By a non-smoothness, corresponding to an actual control function $u(x)$, *we mean the limit of Δ-non-smoothness when $\Delta \rightarrow 0$.*

Defuzzification. For simplicity of analysis, we only use centroid defuzzification.

The formulation of the problem in mathematical terms is now complete. Now, we are ready to describe the main results of this section.

First Stage: Minimizing Relaxation Time (i.e., Maximizing Stability). Let us first describe the result corresponding to the first stage, where we minimize relaxation time.

Theorem 4.4.1. *Assume that an actual control function $u(x)$ is given. Then, among all possible "or"- and "and"-operations, the smallest relaxation time, corresponding to $u(x)$, occurs when we use $f_\vee(a,b) = \min(a+b,1)$ and $f_\&(a,b) = \min(a,b)$.*

Second Stage: Minimizing Non-Smoothness (i.e., Maximizing Smoothness). We have already mentioned that since we are using an "and"-operation for which $f_\&(a,b) = 0$ for some $a,b > 0$, we may end up with a situation where the resulting function $\mu_C(u)$ is identically 0 and therefore, fuzzy control methodology is not applicable. For such a situation, we must restrict ourselves to correlated "and"-operations. For these operations, we get the following result:

Theorem 4.4.2. *Assume that an actual control function $u(x)$ is given. Then among all possible "or"-operations and all possible correlated "and"-operations, the* *smallest non-smoothness, corresponding to* $u(x)$, *occurs when we use* $f_\vee(a,b) = \max(a,b)$ *and* $f_\&(a,b) = a \cdot b$.

General comment. These results are in good accordance with the general optimization results for fuzzy control described in [79]. We show that the optimal pairs of operations described in Theorem 4.4.1 and Theorem 4.4.2 are example of so-called *tropical* (*idempotent*) algebras. Thus, the use of these algebras is indeed a way to optimize fuzzy control.

What Are Tropical Algebras and What Are Idempotent Algebras? In arithmetic, we have two basic operations: addition and multiplication. There are numerous generalizations of these two operations to objects which are more general than numbers: e.g., we can define the sum and (cross) product of two 3D vectors, sum and product of complex numbers, sum and products of matrices, etc. Many results and algorithms originally developed for operations with real numbers have been successfully extended (sometimes, with appropriate modifications) to such more general objects.

It turns out that many of these results can be also extended to the case where one of the operations $\oplus$ is *idempotent*, i.e., where $a \oplus a = a$ for all a. Structures with two related operations one of which is idempotent and another one has the usual properties of addition or multiplication (such as associativity) are called *idempotent algebras*; see, e.g., [71, 88, 89].

The most widely used example of an idempotent algebra is a *tropical algebra*, i.e., an algebra which is isomorphic to a *max-plus* algebra with operations $a \otimes b = a + b$ and $a \oplus b = \max(a,b)$. In precise terms, the set with two operation $f_1(a,b)$ and $f_2(a,b)$ is isomorphic to a max-plus algebra if there is a 1-1 mapping $m(x)$ for which $f_1(a,b)$ get transformed into the sum and $f_2(a,b)$ gets transformed into the maximum, in the sense that $m(f_1(a,b)) = m(a) + m(b)$ and $m(f_2(a,b)) = \max(m(a), m(b))$.

Both Optimal Pairs of "AND"- and "OR"-Operations Form Tropical Algebras. Let us show that – at least until we reach the value 1 – both pairs of optimal "and"- and "or"-operations form tropical algebras, i.e., are isomorphic to the max-plus algebra.

Let us start with operations that maximize stability: $f_\vee(a,b) = \min(a+b,1)$ and $f_\&(a,b) = \min(a,b)$. Until we reach the value 1, we get $f_\vee(a,b) = a+b$ and $f_\&(a,b) = \min(a,b)$. Let us show that the mapping $m(x) = -x$ is the desired isomorphism. Indeed,

$$m(f_\vee(a,b)) = -(a+b) = (-a) + (-b) = m(a) + m(b).$$

Similarly, since the function $m(x) = -x$ is decreasing, it attains its largest value when x is the smallest, in particular, $\max(-a,-b) = -\min(a,b)$. Thus, we have

$$m(f_\&(a,b)) = -\min(a,b) = \max(-a,-b) = \max(m(a), m(b)).$$

So, our two operations are indeed isomorphic to plus and max.

Let us now show that the operations $f_\vee(a,b) = \max(a,b)$ and $f_\&(a,b) = a \cdot b$ that maximize smoothness are also isomorphic to the max-plus algebra. Indeed, in this case, we can take $m(x) = \ln(x)$. Logarithm is an increasing function, so it attains its largest value when x is the largest, in particular, $\max(\ln(a), \ln(b)) = \ln(\max(a,b))$. Thus, we have

$$m(f_\vee(a,b)) = \ln(\max(a,b)) = \max(\ln(a), \ln(b)) = \max(m(a), m(b)).$$

On the other hand, $\ln(a \cdot b) = \ln(a) + \ln(b)$ hence

$$m(f_\&(a,b)) = \ln(a \cdot b) = \ln(a) + \ln(b) = m(a) + m(b).$$

The isomorphism is proven.

Proof of Lemma 4.4.1 is simple, because for small δ the control is approximately linear: $u(x) \approx u'(0) \cdot x$.

Proof of Theorem 4.4.1. Let us first consider the case where $u(x)$ is a linear function i.e., where $u(x) = -k \cdot x$. In this case, instead of directly proving the statement of Theorem 4.4.1 (that the limit of Δ-relaxation times is the biggest for the chosen reasoning method), we prove that for every Δ, Δ-relaxation time is the largest for this very pair of "or"- and "and"-operations. The statement itself can then be easily obtained by turning to a limit $\Delta \to 0$.

So, let us consider the case where $u(x) = -k \cdot x$ for some $k > 0$. In view of Lemma 4.4.1, we must compute the derivative $\bar{u}'(0) = \lim_{x \to 0} (\bar{u}(x) - \bar{u}(0))/x)$, where $\bar{u}(x)$ is the control strategy into which the described fuzzy control methodology translates our rules.

It is easy to show that $\bar{u}(0) = 0$. Hence, $\bar{u}'(0) = \lim \bar{u}(x)/x$. So, to find the desired derivative, we must estimate $\bar{u}(x)$ for small x. To get the limit, it is sufficient to consider only negative values $x \to 0$. Therefore, for simplicity of considerations, let us restrict ourselves to small negative values x (we could as well restrict ourselves to positive x, but we have chosen negative ones because for them the control is positive and therefore, slightly easier to handle).

In particular, we can always take all these x from an interval $[-\Delta/2, 0]$. For such x, only two of the membership functions N_j are different from 0: $N(x) = N_0(x) = 1 - |x|/\Delta$ and $SN(x) = N_{-1}(x) = |x|/\Delta$. Therefore, only two rules are fired for such x, namely, those that correspond to $N(u)$ and $SP(u)$.

We have assumed the centroid defuzzification rule, according to which $\bar{u}(x) = n(x)/d(x)$, where the numerator $n(x) = \int u \cdot \mu_C(u)\, du$ and the denominator is equal to $d(x) = \int \mu_C(u)\, du$. When $x = 0$, the only rule that is applicable is $N_0(x) \to N_0(u)$. Therefore, for this x, the above-given general expression for $\mu_C(u)$ turns into $\mu_C(x) = \mu_N(u)$ Indeed, from our definitions of "and"- and "or"-operations, we can deduce the following formulas:

- $f_\&(a,0) = 0$ for an arbitrary a, so the rule whose condition is not satisfied leads to 0, and
- $f_\vee(a,0) = 0$ for all a, so the rule that leads to 0, does not influence $\mu_C(u)$.

Therefore, for $x = 0$, the denominator $d(0)$ equals $\int \mu_N(u)\, du = k \cdot \Delta$ (this is the area of the triangle that is the graph of the membership function).

So, when $x \to 0$, then $d(x) \to d(0) = k \cdot \Delta$. Therefore, we can simplify the expression for the desired value $\bar{u}'(0)$:

$$\bar{u}'(0) = \lim \frac{u(x)}{x} = \lim \left(\frac{n(x)}{d(x)} \right) / x = (k \cdot \Delta)^{-1} \lim \frac{n(x)}{x}.$$

Since $k\Delta$ is a constant that does not depend on the choice of a reasoning method (i.e., of "or"- and "and"-operations), the biggest value of $\bar{u}'(0)$ (and hence, the smallest relaxation time) is attained when the limit $\lim(n(x)/x)$ takes the smallest possible value. So, from now on, let's estimate this limit.

For small negative x, as we have already mentioned, only two rules are fired: $N(x) \to N(u)$ and $SN(x) \to SP(u)$. Therefore, the membership function for control takes the following form: $\mu_C(u) = f_\vee(p_1(u), p_2(u))$, where $p_1(u) = f_\&(\mu_N(x), \mu_N(u))$ and $p_2(u) = f_\&(\mu_{SN}(x), \mu_{SP}(u))$. The function $\mu_{SP}(u)$ is different from 0 only for $u > 0$. Therefore, for $u < 0$, we have $p_2(u) = 0$ and hence, $\mu_C(u) = p_1(u)$.

We are looking for the reasoning method, for which $\lim(n(x)/x)$ takes the largest possible value, where $n(x) = \int \mu_C(u)\, du$. Let's fix an arbitrary "and"-operation $f_\&$ and consider different functions $f_\vee$. If we use two different "or"-operations $f_\vee(a, b)$ and $g_\vee(a, b)$ for which $f_\vee(a, b) \le g_\vee(a, b)$ for all a, b, then, when we switch from $f_\vee$ to $g_\vee$, the values of $\mu_C(u)$ for $u < 0$ are unaffected, but the values for $u > 0$ increase. Therefore, the total value of the numerator integral $n(x) = \int \mu_C(u)\, du$ increases after this change. So, if we change $f_\vee$ to a maximum possible function $\min(a+b, 1)$, we increase this integral. Therefore, we arrive at a new pair of functions, for which the new value of $\bar{u}$ is not smaller for small x, and, therefore, the derivative of $\bar{u}$ in 0 is not smaller.

Therefore, when looking for the best reasoning methods, it is sufficient to consider only the pairs of "or"- and "and"-operations in which $f_\vee(a, b) = \min(a+b, 1)$. In this case, we have $\mu_C(x) = p_1(u) + p_2(u) - p_{ab}(u)$, where $p_{ab}(u)$ is different from 0 only for $u \approx 0$, and corresponds to the values u for which we use the 1 part of the $\min(a+b, 1)$ formula. Therefore, $n(x)$ can be represented as the sum of the three integrals: $n(x) = n_1 + n_2 - n_{ab}$, where $n_1 = \int u \cdot p_1(u)\, du$, $n_2 = \int u \cdot p_2(u)\, du$, and $n_{ab} = \int u \cdot p_{ab}(u)\, du$. Let's analyze these three components one by one.

- The function $p_1(u)$ is even (because $\mu_N(u)$ is even). It is well known that for an arbitrary even function f, the integral $\int u \cdot f(u)\, du$ equals 0. Therefore, $n_1 = 0$. So, this component does not influence the limit $\lim(n(x)/x)$ (and therefore, does influence the relaxation time).
- The difference $p_{ab}(u)$ is of size u, which, in its turn, is of size x ($p_{ab}(u) \sim u \sim x$), and it is different from 0 on the area surrounding $u = 0$ that is also of size $\sim x$. Therefore, the corresponding integral n_{ab} is of order x^3. Therefore, when $x \to 0$, we have $n_{ab}/x \sim x^2 \to 0$. This means that this component does not influence the limit $\lim(n(x)/x)$ either.

As a result, the desired limit is completely determined by the second component $p_2(u)$, i.e., $\lim \dfrac{n(x)}{x} = \lim \dfrac{n_2(x)}{x}$. Therefore, the relaxation time is the smallest when $\lim \dfrac{n_2(x)}{x}$ takes the biggest possible value. Now,

$$n_2 = \int u \cdot p_2(u) \, du,$$

where $p_2(u) = f_\&(\mu_{SN}(x), \mu_{SP}(u))$. The membership function $\mu_{SP}(u)$ is different from 0 only for positive u. Therefore, the function $p_2(u)$ is different from 0 only for positive u. So, the bigger $f_\&$, the bigger n_2. Therefore, the maximum is attained when $f_\&$ attains its maximal possible value, i.e., the value $\min(a, b)$. For linear actual control functions, the statement of the theorem is thus proven.

The general case follows from the fact that the relaxation time is uniquely determined by the behavior of a system near $x = 0$. The smaller Δ we take, the closer $u(x)$ to a linear function on an interval $[-\Delta, \Delta]$ that determines the derivative of $\bar{u}(x)$, and, therefore, the closer the corresponding relaxation time to a relaxation time of a system that originated from the linear control. Since for each of these approximating systems, the resulting relaxation time is the smallest for a given pair of "or"- and "and"-operations, the same inequality is true for the original system that these linear systems approximate. Q.E.D.

Proof of Theorem 4.4.2. For a linear system $u(x) = -k \cdot x$, we have

$$x(t) = \delta \cdot \exp(-k \cdot t),$$

so $\dot{x}(t) = -k \cdot \delta \cdot \exp(-k \cdot t)$, and the non-smoothness is equal to

$$I(\delta) = \delta^2 \cdot \int_0^\infty k^2 \cdot \exp(-2k \cdot t) \, dt = (k/2) \cdot \delta^2.$$

Therefore, $I = k/2$. For non-linear systems with a smooth control $u(x)$ we can similarly prove that $I = -(1/2) \cdot u'(0)$. Therefore, the problem of choosing a control with the smallest value of non-smoothness is equivalent to the problem of finding a control with the smallest value of $k = |u'(0)|$. This problem is directly opposite to the problem that we solved in Theorem 4.4.1, where our main goal was to maximize k.

Similar arguments show that the smallest value of k is attained when we take the smallest possible function for "or" and the smallest possible operation for "and". Q.E.D.

Comment. We have proved our results only for the simplified plant. However, as one can easily see from the proof, we did not use much of the details about this plant. What we mainly used was the inequalities between different "and"- and "or"-operations. In particular, our proofs do not use the triangular form of the membership function, they use only the fact that the membership functions are located on the intervals $[a - \Delta, a + \Delta]$.

Therefore, a similar proof can be applied in a much more general context. We did not formulate our results in this more general context because we did not want to cloud our results with lots of inevitable technical details.

4.5 How to Combine Different Fuzzy Estimates: A New Justification for Weighted Average Aggregation in Fuzzy Techniques

In Many Practical Situations, We Need to Decide Whether to Accept or to Continue Improving. In many practical situations, we want to have a good solution, so we start with some solution and keep improving it until we decide that this solution is good enough.

For example, this is how software is designed: we design the first version, test it, if the results are satisfactory, we release it, otherwise, if this version still has too many bugs, we continue improving it. Similarly, when a legislature works on a law (e.g., on an annual state budget), it starts with some draft version. If the majority of the legislators believe that this budget is good enough, the budget is approved, otherwise, the members of the legislature continue working on it until the majority is satisfied. Yet another example is home remodeling: the owners hire a company, the company produces a remodeling plan. If the owners are satisfied with this plan, the remodeling starts, if not, the remodeling company makes changes and adjustments until the owners are satisfied.

In Many Such Situations, We Only Have Fuzzy Evaluations of the Solution's Quality. In some cases, the requirements are precisely formulated. For example, for software whose objective is to control critical systems such as nuclear power plants or airplanes, we usually have very precise specifications, and we do not release the software until we are 100% sure that the software satisfies all these specifications.

However, in most other situations, the degree of satisfaction is determined subjectively. Usually, there are several criteria that we want the solution to satisfy. For example, the budget must not contain too many cuts in important services, not contain drastic tax increases, be fair to different parts of the population and to different geographic areas. In many situations, these criteria are not precise, so the only way to decide to what extent each of these criteria is satisfied it to ask experts.

It is natural to describe the experts' degree of satisfaction in each criterion by a real number from the interval $[0, 1]$ so that 0 means no satisfaction at all, 1 means perfect satisfaction, and intermediate values mean partial satisfaction. This is exactly what fuzzy techniques start with.

Many methods are known to elicit the corresponding values from the experts; see, e.g., [70]. For example, if each expert is absolutely confident about whether the given solution satisfies the given criterion or not, we can take, as degree of satisfaction, the proportion of experts who considers this solution satisfactory. For example, if 60% of the experts considers the given aspect of the solution to be satisfactory, then we say that the expert's degree of satisfaction is 0.6. This is how decisions are usually made in legislatures.

In many practical situations, however, experts are not that confident; each expert, instead of claiming that the solution is absolutely satisfactory or absolutely unsatisfactory, feels much more comfortable marking his or her degree of satisfaction on a scale – e.g., on a scale from 0 to 5. This is, e.g., how in the US universities, students evaluate their professors. If a student marks 4 on a scale from 0 to 5 as an answer to the question "Is a professor well organized?", then we can say that the student's degree of satisfaction with the professor's organization of the class is $4/5 = 0.8$. The degrees corresponding to several students are then averaged to form the class evaluation. Similarly, in general, the experts' estimates are averaged.

Formulation of the Problem. Let us assume that we have several (n) criteria. For a given solution, for each of these criteria, we ask the experts and get a degree a_i to which – according to the experts — this particular criterion is satisfied. We need to define a criterion that enables us, based on these n numbers $a_1, \ldots, a_n \in [0, 1]$, to decide whether solution as a whole is satisfactory to us.

In this section, we show that a natural logical symmetry – between true and false values – leads to a reasonable way of combining such expert decisions.

Comment. This result first appeared in [101].

Towards a Formal Description of the Problem. In order to find a reasonable solution to this problem, let us formulate this problem in precise terms.

We need to divide the unit cube $[0, 1]^n$ – the set of all possible values of the tuple $a = (a_1, \ldots, a_n)$ – into two complimentary sets: the set S of all the tuples for which the solution is accepted as satisfactory, and the set U of all the tuples for which the solution is rejected as unsatisfactory.

Natural Requirements. Let us assume that we have two groups of experts whose tuples are a and b, and that, according to both tuples, we conclude that the solution is satisfactory, i.e., that $a \in S$ and $b \in S$. It is then reasonable to require that if we simply place these two groups of experts together, we can still get a satisfactory decision.

Similarly, it is reasonable to conclude that if two groups decide that the solution is unsatisfactory, then by combining their estimates, we should still be able to conclude that the solution is unsatisfactory.

According to our description, when we have two groups of experts consisting of n_a and n_b folks, then, to form a joint tuple, we combine the original tuples with the weights proportional to these numbers, i.e., we consider the tuple

$$c = \frac{n_a}{n_a + n_b} \cdot a + \frac{n_b}{n_a + n_b} \cdot b.$$

Thus, we conclude that if $a, b \in S$ and $r \in [0, 1]$ is a rational number, then

$$r \cdot a + (1 - r) \cdot b \in S.$$

It is also reasonable to require that if, instead of simply averaging, we use arbitrary weights to take into account that some experts are more credible, we should also be able to conclude that the combined group of experts should lead to a satisfactory decision. In other words, we conclude that if $a, b \in S$ and r is an arbitrary number from the interval $[0, 1]$ is a rational number, then we should have $r \cdot a + (1 - r) \cdot b \in S$. In mathematical terms, this means that the set S is *convex*.

Similarly, if $a, b \in U$ and $r \in [0, 1]$, then $r \cdot a + (1 - r) \cdot b \in U$. Thus, the complement U to the set S should be convex.

Analysis of the Requirement. Two disjoint convex sets can always be separated by a half-plane; see, e.g., [138]. In this case, all the satisfactory tuples are on one side, all unsatisfactory points are on the other side. A general hyper-plane can be described by linear equations $\sum w_i \cdot x_i = t$, so all the S-points correspond to $\sum w_i \cdot x_i \geq t$ and all the U points to $\sum w_i \cdot x_i \leq t$,

Conclusion. We have shown that reasonable conditions on decision making indeed leads to the weighted average.

Chapter 5
Possible Ideas for Future Work

In this book, on numerous examples, we showed that symmetries and similarities can help with algorithmic aspects of analysis, prediction, and control in science and engineering. The breadth and depth of these examples show that the approach based on symmetry and similarity is indeed very promising.

However, to make this approach more widely used, additional work is needed. Indeed, in each of our examples, the main challenge is finding the relevant symmetries. As of now, we have found these symmetries on a case-by-case basis, by consulting with the corresponding experts. It would be great to generalize our experience – and experience of other researchers who used symmetry approach – and develop a general methodology of finding the relevant symmetries and similarities. Such a general methodology would help scientists and engineers to apply the promising symmetry- and similarity-based approach to important new practical problems.

© Springer-Verlag Berlin Heidelberg 2015
J. Nava and V. Kreinovich, *Algorithmic Aspects of Analysis, Prediction, and Control in Science
and Engineering*, Studies in Systems, Decision and Control 14, DOI: 10.1007/978-3-662-44955-4_5

References

1. Aczel, J.: Lectures on Functional Differential Equations and their Applications. Dover, New York (2006)
2. Apolloni, B., Bassis, S., Valerio, L.: A moving agent metaphor to model some motions of the brain actors. In: Abstracts of the Conference, Evolution in Communication and Neural Processing from First Organisms and Plants to Man ... and Beyond, Modena, Italy, November 18-19, p. 17 (2010)
3. Averill, M.G.: A Lithospheric Investigation of the Southern Rio Grande Rift. PhD Dissertation, University of Texas at El Paso, Department of Geological Sciences (2007)
4. Averill, M.G., Miller, K.C., Keller, G.R., Kreinovich, V., Araiza, R., Starks, S.A.: Using expert knowledge in solving the seismic inverse problem. International Journal of Approximate Reasoning 45(3), 564–587 (2007)
5. Bardossy, G., Fodor, J.: Evaluation of Uncertainties and Risks in Geology. Springer, Berlin (2004)
6. Becher, V., Heiber, P.A.: A better complexity of finite sequences. In: Abstracts of the 8th Int'l Conf. on Computability and Complexity in Analysis CCA 2011 and 6th International Conf. on Computability, Complexity, and Randomness CCR 2011, Cape Town, South Africa, January 31-February 4, p. 7 (2011)
7. Becher, V., Heiber, P.A.: A linearly computable measure of string complexity. Theoretical Computer Science (to appear)
8. Bedregal, B.C., Reiser, R.H.S., Dimuro, G.P.: Xor-implications and e-implications: classes of fuzzy implications based on fuzzy xor. Electronic Notes in Theoretical Computer Science (ENTCS) 247, 5–18 (2009)
9. Beliakov, G., Pradera, A., Calvo, T.: Aggregation Functions: A Guide for Practitioners. STUDFUZZ, vol. 221. Springer, Heidelberg (2007)
10. Berz, M., Hoffstätter, G.: Computation and Application of Taylor Polynomials with Interval Remainder Bounds. Reliable Computing 4, 83–97 (1998)
11. Berz, M., Makino, K.: Verified Integration of ODEs and Flows using Differential Algebraic Methods on High-Order Taylor Models. Reliable Computing 4, 361–369 (1998)
12. Berz, M., Makino, K., Hoefkens, J.: Verified Integration of Dynamics in the Solar System. Nonlinear Analysis: Theory, Methods, & Applications 47, 179–190 (2001)
13. Bishop, C.M.: Pattern Recognition and Machine Learning. Springer, New York (2007)
14. Blin, J.M.: Fuzzy relations in group decision theory. Journal of Cybernetics 4, 17–22 (1974)
15. Blin, J.M., Whinston, A.B.: Fuzzy sets and social choice. Journal of Cybernetics 3, 28–36 (1973)

16. Bonola, R.: Non-Euclidean Geometry. Dover, New York (2010)
17. Brading, K., Castellani, E. (eds.): Symmetries in Physics: Philosophical Reflections. Cambridge University Press, Cambridge (2003)
18. Branden, C.I., Tooze, J.: Introduction to Protein Structure. Garland Publ., New York (1999)
19. Bravo, S., Nava, J.: Mamdani approach to fuzzy control, logical approach, what else? In: Proceedings of the 30th Annual Conference of the North American Fuzzy Information Processing Society NAFIPS 2011, El Paso, Texas, March 18-20 (2011)
20. Buchanan, B.G., Shortliffe, E.H.: Rule-Based Expert Systems. The Mycin Experiments of the Stanford Heuristic Programming Project. Addison-Wesley, Reading (1984)
21. Bugajski, S.: Delinearization of quantum logic. International Journal of Theoretical Physics 32(3), 389–398 (1993)
22. Cai, L.Y., Kwan, H.K.: Fuzzy Classifications Using Fuzzy Inference Networks. IEEE Transactions on Systems, Man, and Cybernetics. Part B: Cybernetics 28(3), 334–347 (1998)
23. Chang, S.S.L., Zadeh, L.A.: On fuzzy mapping and control. IEEE Transactions on Systems, Man and Cybernetics SMC-2, 30–34 (1972)
24. Cormen, T.H., Leiserson, C.E., Rivest, R.L., Stein, C.: Introduction to Algorithms. MIT Press, Cambridge (2009)
25. Davies, E.B., Lewis, J.T.: An operational approach to quantum probability. Communications in Mathematical Physics 17(3), 239–260 (1970)
26. D'Azzo, J.J., Houpis, C.H.: Linear Control System Analysis And Design: Conventional And Modern. Mc-Graw Hill, NY (1988)
27. Demicco, R., Klir, G. (eds.): Fuzzy Logic in Geology. Academic Press (2003)
28. Doser, D.I., Crain, K.D., Baker, M.R., Kreinovich, V., Gerstenberger, M.C.: Estimating uncertainties for geophysical tomography. Reliable Computing 4(3), 241–268 (1998)
29. Došlić, T., Klein, D.J.: Splinoid interpolation on finite posets. Journal of Computational and Applied Mathematics 177, 175–185 (2005)
30. Dowding, K., Pilch, M., Hills, R.: Formulation of the thermal problem. Computer Methods in Applied Mechanics and Engineering 197(29-32), 2385–2389 (2008)
31. Dravskikh, A., Finkelstein, A., Kreinovich, V.: Astrometric and geodetic applications of VLBI 'arc method'. In: Proceedings of the IAU Colloquium Modern Astrometry, Vienna, vol. 48, pp. 143–153 (1978)
32. Dravskikh, A., Kosheleva, O., Kreinovich, V., Finkelstein, A.: The method of arcs and differential astrometry. Soviet Astronomy Letters 5(3), 160–162 (1979)
33. Dubois, D., Prade, H.: Fuzzy Sets and Systems: Theory And Applications. Academic Press, NY (1980)
34. Dubois, D., Prade, H.: Possibility Theory. An Approach to Computerized Processing of Uncertainty. Plenum Press, NY (1988)
35. Elliot, J.P., Dawber, P.G.: Symmetry in Physics, vol. 1. Oxford University Press, New York (1979)
36. Evans, L.C.: Partial Differential Equations. American Mathematical Society, Providence (1998)
37. Ferson, S., Oberkampf, W., Ginzburg, L.: Model validation and predictive capacity for the thermal challenge problem. Computer Methods in Applied Mechanics and Engineering 197(29-32), 2408–2430 (2008)
38. Feynman, R., Leighton, R., Sands, M.: The Feynman Lectures on Physics. Addison Wesley, Boston (2005)
39. Finkelstein, A., Kosheleva, O., Kreinovich, V.: Astrogeometry, error estimation, and other applications of set-valued analysis. ACM SIGNUM Newsletter 31(4), 3–25 (1996)

40. Finkelstein, A., Kosheleva, O., Kreinovich, V.: Astrogeometry: towards mathematical foundations. International Journal of Theoretical Physics 36(4), 1009–1020 (1997)
41. Finkelstein, A., Kosheleva, O., Kreinovich, V.: Astrogeometry: geometry explains shapes of celestial bodies. Geombinatorics VI(4), 125–139 (1997)
42. Finkelstein, A.M., Kreinovich, V., Zapatrin, R.R.: Fundamental physical equations are uniquely determined by their symmetry groups. Springer Lecture Notes on Mathematics, vol. 1214, pp. 159–170 (1986)
43. Finkelstein, A.M., Kreinovich, V.: Derivation of Einstein's, Brans-Dicke and other equations from group considerations. In: Choque-Bruhat, Y., Karade, T.M. (eds.) On Relativity Theory. Proceedings of the Sir Arthur Eddington Centenary Symposium, Nagpur, India, vol. 2, pp. 138–146. World Scientific, Singapore (1984)
44. Foulis, D.J., Bennett, M.K.: Effect algebras and unsharp quantum logics. Foundations of Physics 24(10), 1331–1352 (1994)
45. Garloff, J.: The Bernstein algorithm. Interval Computation 2, 154–168 (1993)
46. Garloff, J.: The Bernstein expansion and its applications. Journal of the American Romanian Academy 25-27, 80–85 (2003)
47. Garloff, J., Graf, B.: Solving strict polynomial inequalities by Bernstein expansion. In: Munro, N. (ed.) The Use of Symbolic Methods in Control System Analysis and Design, London. IEE Contr. Eng., vol. 56, pp. 339–352 (1999)
48. Garloff, J., Smith, A.P.: Solution of systems of polynomial equations by using Bernstein polynomials. In: Alefeld, G., Rohn, J., Rump, S., Yamamoto, T. (eds.) Symbolic Algebraic Methods and Verification Methods – Theory and Application, pp. 87–97. Springer, Wien (2001)
49. Gaubert, S., Katz, R.D.: The tropical analogue of polar cones. Linear Algebra and Appl. 431(5-7), 608–625 (2009)
50. Gelfand, I.M., Fomin, S.V.: Calculus of Variations. Dover Publ., New York (2000)
51. Götze, F., Tikhomirov, A.: Rate of convergence in probability to the Marchenko-Pastur law. Bernoulli 10(3), 503–548 (2004)
52. Gromov, M.: Crystals, proteins and isoperimetry. Bulletin of the American Mathematical Society 48(2), 229–257 (2011)
53. Hamhalter, J.: Quantum structures and operator algebras. In: Engesser, K., Gabbay, D.M., Lehmann, D. (eds.) Handbook of Quantum Logic and Quantum Structures: Quantum Structures, pp. 285–333. Elsevier, Amsterdam (2007)
54. Hennessy, J.L., Patterson, D.A.: Computer Architecture: A Quantitative Approach. Morgan Kaufmann, San Francisco (2007)
55. Hernandez, J.E., Nava, J.: Least sensitive (most robust) fuzzy 'exclusive or' operations. In: Proceedings of the 30th Annual Conference of the North American Fuzzy Information Processing Society NAFIPS 2011, El Paso, Texas, March 18-20 (2011)
56. Hersch, H.M., Caramazza, A.: A fuzzy-set approach to modifiers and vagueness in natural languages. Journal of Experimental Psychology: General 105, 254–276 (1976)
57. Hills, R., Dowding, K., Swiler, L.: Thermal challenge problem: summary. Computer Methods in Applied Mechanics and Engineering 197(29-32), 2490–2495 (2008)
58. Hills, R., Pilch, M., Dowding, K., Red-Horse, J., Paez, T., Babuska, I., Tempone, R.: Validation Challenge Workshop. Computer Methods in Applied Mechanics and Engineering 197(29-32), 2375–2380 (2008)
59. Hoefkens, J., Berz, M.: Verification of invertibility of complicated functions over large domains. Reliable Computing 8(1), 1–16 (2002)
60. Hole, J.A.: Nonlinear high-resolution three-dimensional seismic travel time tomography. Journal of Geophysical Research 97(B5), 6553–6562 (1992)

61. Ivanciuc, T., Klein, D.J.: Parameter-free structure-property correlation via progressive reaction posets for substituted benzenes. Journal of Chemical Information and Computer Sciences 44(2), 610–617 (2004)

62. Ivanciuc, T., Ivanciuc, O., Klein, D.J.: Posetic quantitative superstructure/activity relationships (QSSARs) for chlorobenzenes. Journal of Chemical Information and Modeling 45, 870–879 (2005)

63. Ivanciuc, T., Ivanciuc, O., Klein, D.J.: Modeling the bioconcentration factors and bioaccumulation factors of polychlorinated biphenyls with posetic quantitative superstructure/activity relationships (QSSAR). Molecular Diversity 10, 133–145 (2006)

64. Ivanciuc, T., Klein, D.J., Ivanciuc, O.: Posetic cluster expansion for substitution-reaction networks and application to methylated cyclobutanes. Journal of Mathematical Chemistry 41(4), 355–379 (2007)

65. Kissling, E.: Geotomography with local earthquake data. Reviews of Geophysics 26(4), 659–698 (1998)

66. Kissling, E., Ellsworth, W., Cockerham, R.S.: 3-D structure of the Long Valley, California, region by tomography. In: Proceedings of the XIX Workshop on Active Tectonic and Magmatic Processes Beneath Long Valley Caldera, Eastern California. U.S. Geological Survey, Open-File Report 84-939, pp. 188–220 (1984)

67. Klein, D.J.: Chemical graph-theoretic cluster expansions. International Journal of Quantum Chemistry, Quantum Chemistry Symposium 20, 153–171 (1986)

68. Klein, D.J., Bytautas, L.: Directed reaction graphs as posets. Communications in Mathematical and in Computer Chemistry (MATCH) 42, 261–290 (2000)

69. Klir, G.J., Folger, T.A.: Fuzzy Sets, Uncertainty and Information. Prentice Hall, Englewood Cliffs (1988)

70. Klir, G., Yuan, B.: Fuzzy Sets and Fuzzy Logic: Theory and Applications. Prentice Hall, Upper Saddle River (1995)

71. Kolokoltsov, V.N., Maslov, V.P.: Idempotent Analysis and Its Applications. Springer, Heidelberg (2010)

72. Kosko, B.: Neural Networks and Fuzzy Systems. Prentice Hall, Englewood Cliffs (1992)

73. Kreinovich, V., Mouzouris, G.C., Nguyen, H.T.: Fuzzy rule based modeling as a universal approximation tool. In: Nguyen, H.T., Sugeno, M. (eds.) Fuzzy Systems: Modeling and Control, pp. 135–195. Kluwer, Boston (1998)

74. Kreinovich, V., Nava, J.: I-complexity and discrete derivative of logarithms: a group-theoretic explanation. In: Abstracts of the 9th Joint NMSU/UTEP Workshop on Mathematics, Computer Science, and Computational Sciences, Las Cruces, New Mexico (April 2, 2011)

75. Kreinovich, V., Nava, J.: I-complexity and discrete derivative of logarithms: a symmetry-based explanation. Journal of Uncertain Systems 6(2), 118–121 (2012)

76. Kreinovich, V., Nava, J., Romero, R., Olaya, J., Velasco, A., Miller, K.C.: Spatial resolution for processing seismic data: type-2 methods for finding the relevant granular structure. In: Proceedings of the IEEE International Conference on Granular Computing GrC 2010, Silicon Valley, USA, August 14-16 (2010)

77. Kreinovich, V., Perfilieva, I.: From gauging accuracy of quantity estimates to gauging accuracy and resolution of measuring physical fields. In: Wyrzykowski, R., Dongarra, J., Karczewski, K., Wasniewski, J. (eds.) PPAM 2009, Part II. LNCS, vol. 6068, pp. 456–465. Springer, Heidelberg (2010)

78. Kreinovich, V., Quintana, C.: Neural networks: what non- linearity to choose? In: Fredericton, N.B. (ed.) Proceedings of the 4th University of New Brunswick AI Workshop, Fredericton, N.B., Canada, pp. 627–637 (1991)

79. Kreinovich, V., Quintana, C., Lea, R., Fuentes, O., Lokshin, A., Kumar, S., Boricheva, I., Reznik, L.: What non-linearity to choose? Mathematical foundations of fuzzy control. In: Proceedings of the 1992 International Conference on Fuzzy Systems and Intelligent Control, Louisville, KY, pp. 349–412 (1992)

80. Lea, R.N.: Automated space vehicle control for rendezvous proximity operations. Telemechanics and Informatics 5, 179–185 (1988)

81. Lea, R.N., Jani, Y.K., Berenji, H.: Fuzzy logic controller with reinforcement learning for proximity operations and docking. In: Proceedings of the 5th IEEE International Symposium on Intelligent Control, vol. 2, pp. 903–906 (1990)

82. Lea, R.N., Togai, M., Teichrow, J., Jani, Y.: Fuzzy logic approach to combined translational and rotational control of a spacecraft in proximity of the Space Station. In: Proceedings of the 3rd International Fuzzy Systems Association Congress, pp. 23–29 (1989)

83. Lee, J.M.: Manifolds and Differential Geometry, American Mathematical Society, Providence, Rhode Island (2009)

84. Le Capitaine, H., Frélicot, C.: A new fuzzy 3-rules pattern classifier with reject options based on aggregation of membership degrees. In: Proceedings of the 12th International Conference on Information Processing and Management of Uncertainty in Knowledge-Based Systems IPMU 2008, Malaga, Spain (2008)

85. Lesk, A.M.: Introduction to Protein Science: Architecture, Function, and Genomics. Oxford University Press, New York (2010)

86. Li, S., Ogura, Y., Kreinovich, V.: Limit Theorems and Applications of Set Valued and Fuzzy Valued Random Variables. Kluwer Academic Publishers, Dordrecht (2002)

87. Li, M., Vitanyi, P.: An Introduction to Kolmogorov Complexity and Its Applications. Springer, Heidelberg (2008)

88. Litvinov, G.L., Maslov, V.P. (eds.): Idempotent Mathematics And Mathematical Physics. American Mathematical Society, Providence (2005)

89. Litvinov, G.L., Sergeev, S.N. (eds.): Tropical and Idempotent Mathematics. American Mathematical Society, Providence (2009)

90. Litvinov, G.L., Sobolevskii, A.N.: Idempotent interval analysis and optimization problems. Reliable Computing 7(5), 353–377 (2001)

91. Lohner, R.: Einschliessung der Lösung gewöhnlicher Anfangs- und Randwertaufgaben und Anwendungen. Ph.D. thesis, Universität Karlsruhe, Karlsruhe, Germany (1988)

92. Maceira, M., Taylor, S.R., Ammon, C.J., Yang, X., Velasco, A.A.: High-resolution Rayleigh wave slowness tomography of Central Asia. Journal of Geophysical Research 110, paper B06304 (2005)

93. Mamdani, E.H.: Application of fuzzy algorithms for control of simple dynamic plant. Proceedings of the IEE 121(12), 1585–1588 (1974)

94. Marchenko, V.A., Pastur, L.A.: Distribution of eigenvalues for some sets of random matrices. Matematicheski Sbornik 72(114), 507–536 (1967)

95. Mascarilla, L., Berthier, M., Frélicot, C.: A k-order fuzzy OR operator for pattern classification with k-order ambiguity rejection. Fuzzy Sets and Systems 159(15), 2011–2029 (2008)

96. Moore, R.E., Kearfott, R.B., Cloud, M.J.: Introduction to Interval Analysis. SIAM Press, Philadelphia (2009)

97. Nataraj, P.S.V., Arounassalame, M.: A new subdivision algorithm for the Bernstein polynomial approach to global optimization. International Journal of Automation and Computing 4, 342–352 (2007)

98. Nava, J.: Towards algebraic foundations of algebraic fuzzy logic operations: aiming at the minimal number of requirements. In: Abstracts of the 7th Joint UTEP/NMSU Workshop on Mathematics, Computer Science, and Computational Sciences, Las Cruces, New Mexico (April 3, 2010)

99. Nava, J.: Towards chemical applications of Dempster-Shafer-type approach: case of variant ligands. In: Proceedings of the 30th Annual Conference of the North American Fuzzy Information Processing Society NAFIPS 2011, El Paso, Texas, March 18-20 (2011)

100. Nava, J.: Tropical (idempotent) algebras as a way to optimize fuzzy control. International Journal of Innovative Management, Information & Production (IJIMIP) 2(3) (2011) (to appear)

101. Nava, J.: A new justification for weighted average aggregation in fuzzy techniques. Journal of Uncertain Systems 6(2), 84–85 (2012)

102. Nava, J., Ferret, J., Kreinovich, V., Berumen, G., Griffin, S., Padilla, E.: Why Feynman path integration? Journal of Uncertain Systems 5(2), 102–110 (2011)

103. Nava, J., Kosheleva, O., Kreinovich, V.: Why Bernstein Polynomials Are Better: Fuzzy-Inspired Justification. In: Proceedings of the 2012 IEEE World Congress on Computational Intelligence, WCCI 2012, Brisbane, Australia, June 10-15, pp. 1986–1991 (2012)

104. Nava, J., Kreinovich, V.: Towards interval techniques for model validation. In: Abstracts of the 14th GAMM-IMACS International Symposium on Scientific Computing, Computer Arithmetic and Validated Numerics SCAN 2010, Lyon, France, September 27-30, pp. 99–101 (2010)

105. Nava, J., Kreinovich, V.: Equivalence of Gian-Carlo Rota poset approach and Taylor series approach extended to variant ligands. Journal of Uncertain Systems 5(2), 111–118 (2011)

106. Nava, J., Kreinovich, V.: Orthogonal bases are the best: a theorem justifying Bruno Apolloni's heuristic neural network idea. In: Abstracts of the 9th Joint NMSU/UTEP Workshop on Mathematics, Computer Science, and Computational Sciences, Las Cruces, New Mexico (April 2, 2011)

107. Nava, J., Kreinovich, V.: A Simple Physics-Motivated Equivalent Reformulation of P=NP that Makes This Equality (Slighty) More Plausible, University of Texas at El Paso, Department of Computer Science. Technical Report UTEP-CS-11-45

108. Nava, J., Kreinovich, V.: Theoretical Explanation of Bernstein Polynomials' Efficiency: They Are Optimal Combination of Optimal Endpoint-Related Functions, University of Texas at El Paso, Department of Computer Science. Technical Report UTEP-CS-11-37 (July 2011), http://www.cs.utep.edu/vladik/2011/tr11-37.pdf

109. Nava, J., Kreinovich, V.: Why Neural Networks Are Computationally Efficient Approximators: An Explanation, University of Texas at El Paso, Department of Computer Science. Technical Report UTEP-CS-11-40

110. Nava, J., Kreinovich, V.: Orthogonal bases are the best: a theorem justifying Bruno Apolloni's heuristic neural network idea. Journal of Uncertain Systems 6(2), 122–127 (2012)

111. Nava, J., Kreinovich, V.: Towards interval techniques for model validation. Computing 94(2), 257–269 (2012)

112. Nava, J., Kreinovich, V.: Towards symmetry-based explanation of (approximate) shapes of alpha-helices and beta-sheets (and beta-barrels) in protein structure. Symmetry 4(1), 15–25 (2012)

113. Nava, J., Kreinovich, V., Restrepo, G., Klein, D.J.: Discrete Taylor series as a simple way to predict properties of chemical substances like benzenes and cubanes. Journal of Uncertain Systems 4(4), 270–290 (2010)

114. Neumaier, A.: Taylor forms – use and limits. Reliable Computing 9, 43–79 (2002)

115. Nedialkov, N.S., Kreinovich, V., Starks, S.A.: Interval arithmetic, affine arithmetic, Taylor series methods: why, what next? Numerical Algorithms 37, 325–336 (2004)

116. Nguyen, H.T., Kreinovich, V.: Towards theoretical foundations of soft computing applications. International Journal on Uncertainty, Fuzziness, and Knowledge-Based Systems (IJUFKS) 3(3), 341–373 (1995)
117. Nguyen, H.T., Kreinovich, V., Tolbert, D.: On robustness of fuzzy logics. In: Proceedings of the IEEE International Conference on Fuzzy Systems FUZZ-IEEE 1993, San Francisco, California, vol. 1, pp. 543–547 (March 1993)
118. Nguyen, H.T., Kreinovich, V., Tolbert, D.: A measure of average sensitivity for fuzzy logics. International Journal on Uncertainty, Fuzziness, and Knowledge-Based Systems 2(4), 361–375 (1994)
119. Nguyen, H.T., Kreinovich, V.: Applications of Continuous Mathematics to Computer Science. Kluwer, Dordrecht (1997)
120. Nguyen, H.T., Walker, E.A.: A First Course in Fuzzy Logic. Chapman & Hall/CRC, Boca Raton (2006)
121. Nica, A., Speicher, R.: Lectures on the Combinatorics of Free Probability Theory. Cambridge Univ. Press, Cambridge (2006)
122. Nielsen, M., Chuang, I.: Quantum Computation and Quantum Information. Cambridge University Press, Cambridge (2000)
123. Novotny, J., Bruccoleri, R.E., Newell, J.: Twisted hyperboloid (Strophoid) as a model of beta-barrels in proteins. J. Mol. Biol. 177, 567–573 (1984)
124. Oberkampf, W., Roy, C.: Verification and Validation in Scientific Computing. Cambridge University Press, Cambridge (2010)
125. Oden, G.C.: Integration of fuzzy logical information. Journal of Experimental Psychology: Human Perception Perform 3(4), 565–575 (1977)
126. Osegueda, R., Keller, G., Starks, S.A., Araiza, R., Bizyaev, D., Kreinovich, V.: Towards a general methodology for designing sub-noise measurement procedures. In: Proceedings of the 10th IMEKO TC7 International Symposium on Advances of Measurement Science, St. Petersburg, Russia, June 30-July 2, vol. 1, pp. 59–64 (2004)
127. Parker, R.L.: Geophysical Inverse Theory. Princeton University Press, Princeton (1994)
128. Pedrycz, W., Skowron, A., Kreinovich, V. (eds.): Handbook on Granular Computing. Wiley, Chichester (2008)
129. Pedrycz, W., Succi, G.: FXOR fuzzy logic networks. Soft Computing 7, 115–120 (2002)
130. Perfilieva, I.: Fuzzy transforms: theory and applications. Fuzzy Sets and Systems 157, 993–1023 (2006)
131. Perfilieva, I.: Fuzzy transforms: a challenge to conventional transforms. In: Advances in Imaging and Electron Physics, vol. 47, pp. 137–196 (2007)
132. Perfilieva, I., Kreinovich, V.: A broad prospective on fuzzy transforms: from gauging accuracy of quantity estimates to gauging accuracy and resolution of measuring physical fields. Neural Network World 20(1), 7–25 (2010)
133. Perfilieva, I., Kreinovich, V.: A new universal approximation result for fuzzy systems, which reflects CNF–DNF duality. International Journal of Intelligent Systems 17(12), 1121–1130 (2002)
134. Pilch, M.: Preface: Sandia National Laboratories Validation Challenge Workshop. Computer Methods in Applied Mechanics and Engineering 197(29-32), 2373–2374
135. Pinheiro da Silva, P., Velasco, A., et al.: Propagation and provenance of probabilistic and interval uncertainty in cyberinfrastructure-related data processing and data fusion. In: Muhanna, R.L., Mullen, R.L. (eds.) Proceedings of the International Workshop on Reliable Engineering Computing, REC 2008, Savannah, Georgia, February 20-22, pp. 199–234 (2008)
136. Rabinovich, S.: Measurement Errors and Uncertainties: Theory and Practice. American Institute of Physics, New York (2005)

137. Ray, S., Nataraj, P.S.V.: A new strategy for selecting subdivision point in the Bernstein approach to polynomial optimization. Reliable Computing 14, 117–137 (2010)
138. Rockafellar, R.T.: Convex Analysis. Princeton University Press, Princeton (1996)
139. Rota, G.-C.: On the foundations of combinatorial theory I. Theory of Möbius functions. Zeit. Wahrscheinlichkeitstheorie 2, 340–368 (1964)
140. Sallares, V., Dañobeitia, J.J., Flueh, E.R.: Seismic tomography with local earthquakes in Costa Rica. Tectonophysics 329, 61–78 (2000)
141. Sandoval, S., Kissling, E., Ansorge, J.: High-resolution body wave tomography beneath the SVEKALAPKO array: II. Anomalous upper mantle structure beneath central Baltic Schield. Geophysical Journal International 157, 200–214 (2004)
142. Savage, L.J.: The Foundations of Statistics. Wiley, NY (1954)
143. Sheskin, D.J.: Handbook of Parametric and Nonparametric Statistical Procedures. Chapman & Hall/CRC, Boca Raton (2007)
144. Shortliffe, E.H.: Computer-Based Medical Consultation: MYCIN. Elsevier, NY (1976)
145. Smith, M.H., Kreinovich, V.: Optimal strategy of switching reasoning methods in fuzzy control. In: Nguyen, H.T., Sugeno, M., Tong, R., Yager, R. (eds.) Theoretical Aspects of Fuzzy Control, pp. 117–146. J. Wiley, New York (1995)
146. Sinha, S., Sorkin, R.D.: A sum-over-histories account of an EPR(B) experiment. Foundations of Physics Letters 4, 303–335 (1991)
147. Stec, B., Kreinovich, V.: Geometry of protein structures. I. Why hyperbolic surfaces are a good approximation for beta-sheets. Geombinatorics 15(1), 18–27 (2005)
148. Thompson, A., Moran, J., Swenson Jr., G.: Interferometry and Synthesis in Radio Astronomy. Wiley, New York (2001)
149. Toomey, D.R., Fougler, G.R.: Tomographic inversion of local earthquake data from the Hengill-Grensdalur Central Volcano Comple, Iceland. Journal of Geophysical Research 94(B12), 17,497–17,510 (1989)
150. Tselentis, G.-A., Serpetsidaki, A., Martakis, N., Sokos, E., Paraskevopoulos, P., Kapotas, S.: Local high-resolution passive seismic tomography and Kohonen neural networks – Application at the Rio-Antirio Strait, central Greece. Geophysics 72(4), B93–B106 (2007)
151. Vanderbei, R.: Linear Programming: Foundations and Extensions. Springer, Berlin (2008)
152. Venisti, N., Calcagnile, G., Pontevivo, A., Panza, G.F.: Tomographic study of the Adriatic Plate. Pure and Applied Geophysics 162, 311–329 (2005)
153. Verschuur, G., Kellerman, K.: Galactic and Extragalactic Radio Astronomy. Springer, New York (1988)
154. Wiener, N.: Cybernetics, or Control and Communication in the animal and the machine. MIT Press, Cambridge (1962)
155. Yager, R.R., Liu, L. (eds.): Classic Works of the Dempster-Shafer Theory of Belief Functions. STUDFUZZ, vol. 219. Springer, Heidelberg (2008)
156. Zadeh, L.: Fuzzy sets. Information and Control 8, 338–353 (1965)
157. Zhang, H., Thurber, C.: Development and applications of double-difference seismic tomography. Pure and Applied Geophysics 163, 373–403 (2006)
158. Zimmerman, H.J.: Results of empirical studies in fuzzy set theory. In: Klir, G.J. (ed.) Applied General System Research, pp. 303–312. Plenum, NY (1978)
159. Zimmermann, H.J.: Fuzzy Set Theory and Its Applications. Kluwer, NY (1991)
160. Zollo, A., De Matteis, R., D'Auria, L., Virieux, J.: A 2D non-linear method for travel time tomography: application to Mt. Vesuvius active seismic data. In: Boschi, E., Ekström, G., Morelli, A. (eds.) Problems in Geophysics for the New Millennium, pp. 125–140. Compositori, Bologna (2000)

Index

Printed in the United States
By Bookmasters